Springer Texts in
Electrical Engineering

Consulting Editor: John B. Thomas

Springer Texts in Electrical Engineering

Robert J. Marks II

Introduction to Shannon Sampling and Interpolation Theory

With 109 Illustrations

Springer-Verlag
New York Berlin Heidelberg London
Paris Tokyo Hong Kong Barcelona

Robert J. Marks II
Interactive Systems Design Laboratory
Department of Electrical Engineering
University of Washington
Seattle, Washington 98195
USA

Printed on acid-free paper.

© 1991 Springer-Verlag New York Inc.
Softcover reprint of the hardcover 1st edition 1991

Camera-ready copy prepared by the author using LaTeX.

9 8 7 6 5 4 3 2 1

ISBN-13:978-1-4613-9710-6   e-ISBN-13:978-1-4613-9708-3
DOI: 10.1007/978-1-4613-9708-3

dedicated to:

The Christ Jesus

. . . because of Whom it all makes sense.

Preface

The contents of this book evolved from a set of lecture notes prepared for a graduate survey course on Shannon sampling and interpolation theory. The course was taught at the Department of Electrical Engineering at the University of Washington, Seattle. The book is written for a second-year graduate student who has an established foundation in Fourier analysis and stochastic processes.

Each of the seven chapters in this book includes a list of references specific to that chapter. A sequel to this book will contain an extensive bibliography on the subject. I have also opted to include solutions to selected exercises in the Appendix.

The patience and support of my wife, Monika, and children, Marilee, Joshua and Jeremiah, during the preparation of this book, is gratefully acknowledged.

The input and assistance of numerous students in the formation of this book is also gratefully acknowledged. I am particularly indebted to Payman Arabshahi for his patient and thorough proofreading efforts, J.J. Choi and D.C. Park for their tireless efforts in the preparation of the manuscript, and Seho Oh for his generation of a seemingly endless number of figures and plots.

Contents

1

Introduction

Much of that which is ordinal is modeled as analog. Most computational engines, on the other hand, are digital. Transforming from analog to digital is straightforward: we simply sample. Regaining the original signal from these samples or assessing the information lost in the sampling process are the fundamental questions addressed by sampling and interpolation theory.

This book deals with understanding, generalizing and extending the cardinal series of Shannon sampling theory. The fundamental form of this series states, remarkably, that a bandlimited signal is uniquely specified by its sufficiently close equally spaced samples.

The cardinal series has many names, including the *Whittaker-Shannon sampling theorem* [Goodman (1968)], the *Whittaker-Shannon-Kotel'nikov sampling theorem* [Jerri (1977)], and the *Whittaker-Shannon-Kotel'nikov-Kramer sampling theorem* [Jerri (1977)]. For brevity, we will use the terms *sampling theorem* and *cardinal series.*

1.1   The Cardinal Series

If a signal has finite energy, thc minimum sampling rate is equal to two samples per period of the highest frequency component of the signal. Specifically, if the highest frequency component of the signal is B hertz, then the signal, $x(t)$, can be recovered from the samples by

$$x(t) = \frac{1}{\pi} \sum_{n=-\infty}^{\infty} x\left(\frac{n}{2B}\right) \frac{\sin[\pi(2Bt - n)]}{2Bt - n}.$$

The frequency B is also referred to as the signal's bandwidth and, if B is finite, $x(t)$ is said to be bandlimited [Slepian (1976)].

1.2 History

The history of Shannon sampling theory and the cardinal series is intriguing. A summary of key events in the development of the cardinal series is listed in Table 1.1.

H.S. Black credits Cauchy for recognition of the mechanics of band-limited signal sampling in 1841 and even offers the following translation from Cauchy's original French text:

> IF A SIGNAL IS A MAGNITUDE-TIME FUNCTION, AND IF TIME IS DIVIDED INTO EQUAL INTERVALS SUCH THAT EACH SUBDIVISION COMPRISES AN INTERVAL T SECONDS LONG WHERE T IS LESS THAN HALF THE PERIOD OF THE HIGHEST SIGNIFICANT FREQUENCY COMPONENT OF THE SIGNAL, AND IF ONE INSTANTANEOUS SAMPLE IS TAKEN FROM EACH SUBINTERVAL IN ANY MANNER; THEN A KNOWLEDGE OF THE INSTANTANEOUS MAGNITUDE OF EACH SAMPLE PLUS A KNOWLEDGE OF THE INSTANT WITHIN EACH SUBINTERVAL AT WHICH THE SAMPLE IS TAKEN CONTAINS ALL THE INFORMATION OF THE ORIGINAL SIGNAL.

In a later historical overview, Higgins (1985), however, notes that such a quote was not included in the paper by Cauchy that was cited by Black. Higgins, rather, credits E. Borel in 1897 for the initial recognition of the cardinal series and cites the following passage translated from the original French:

> CONSIDER

$$f(z) = \int_{-\pi}^{\pi} \Psi(x) \, e^{jzx} dx$$

> AND SUPPOSE THAT THE FUNCTION $\Psi(x)$ SATISFIES THE CONDITIONS OF DIRICHLET. IF ONE

KNOWS THE VALUES OF THE FUNCTION, $f(z)$, AT
THE POINTS $z = 0, \pm 1, \pm 2, \ldots$, THEN THE FUNC-
TION $\Psi(x)$ IS COMPLETELY DETERMINED AND,
CONSEQUENTLY, THE ENTIRE FUNCTION $f(z)$ IS
KNOWN WITHOUT AMBIGUITY.

This connection of the Fourier series to the sampling theorem
was the same tool of explanation later used by Shannon in his
classic paper.

E. T. Whittaker published his highly cited paper on the sampling theorem in 1915. In his work, if one function had the same
uniformly spaced samples as another, the functions were said
to be *cotabular*. The sampling theorem interpolation from these
samples resulted in what Whittaker called the *cardinal function*.
The interpolation formula was later dubbed the *cardinal series*
by Whittaker's son, J. M. Whittaker. Among other things, the
senior Whittaker showed the functions, $x(t)$ to which the cardinal series applied were bandlimited and entire on the finite t
plane. He also noted that applicability of the cardinal series to
a function was independent of the choice of sampling phase.

The sampling theorem was reported in the Soviet literature in
a paper by Kotel'nikov in 1933. Shannon (1948) used the sampling theorem to demonstrate that an analog bandlimited signal
was equivalent in an information sense to the series of its samples
taken at the Nyquist rate. He was aware of the work of Whittaker which he cited. Other noted historical generalizations and
extensions of the sampling theorem are listed chronologically in
Table 1.1.

1841	Cauchy's recognition of the Nyquist rate.[1]
1897	Borel's recognition of the feasibility of regaining a bandlimited signal from its samples.
1915	E.T. Whittaker publishes his highly cited paper on the cardinal series.
1928	H. Nyquist establishes the time–bandwidth product of a signal.
1929	J.M. Whittaker coins the term *cardinal series*.
1933	A. Kotel'nikov publishes the sampling theorem in the Soviet literature.
1948	C.E. Shannon publishes a paper which establishes the field of information theory. The sampling theorem is included.
1959	H.P. Kramer generalizes the sampling theorem to functions that are bandlimited in other than the Fourier sense.
1962	D.P. Peterson and D. Middleton extend the sampling theorem to higher dimensions.
1968	A. Papoulis first publishes his generalization of the sampling theorem. A number of previously published extensions are shown to be special cases.

Table 1.1: Key events in the development of the cardinal series.

---

[1] Disputed. See text.

REFERENCES

H.S. Black, **Modulation Theory**, Van Nostrand, New York, 1953.

E. Borel "Sur l'interpolation", *C.R. Acad. Sci. Paris*, Vol. 124, pp. 673-676 (1897).

J.W. Goodman, **Introduction to Fourier Optics**, McGraw-Hill, New York, 1968.

J.R. Higgins "Five short stories about the cardinal series", *Bull. Am. Math. Soc.*, vol. 12, pp.45-89 (1985).

A.J. Jerri "The Shannon sampling theorem - its various extension and applications: a tutorial review", *Proc. IEEE*, vol. 65, pp.1565-1596 (1977).

V.A. Kotel'nikov "On the transmission capacity of "ether" and wire in electrocommunications", *Izd. Red. Upr. Svyazi RKKA* (Moscow), 1933.

H.P. Kramer "A generalized sampling theorem", *J. Math. Phys.*, vol. 38, pp.68-72 (1959).

H. Nyquist "Certain topics in telegraph transmission theory", *AIEE Trans.*, vol. 47, pp.617-644 (1928).

D.P. Petersen and D. Middleton "Sampling and reconstruction of wave number limited functions in N dimensional Euclidean spaces", *Information Contr.*, vol. 5, pp. 279-323 (1962).

A. Papoulis, **Systems and Transforms with Applications in Optics**, McGraw-Hill, new York, 1968.

A. Papoulis "A new algorithm in spectral analysis and bandlimited signal extrapolation", *IEEE Transactions on Circuits and Systems*, vol. CAS-22, pp.735-742 (1975).

A. Papoulis, **Signal Analysis**, McGraw-Hill, New York, 1977.

A. Papoulis "Generalized sampling expansion", *IEEE Transactions on Circuits and Systems*, vol. CAS-24, pp. 652-654 (1977).

C. Shannon "A mathematical theory of communication", *Bell System Technical Journal*, vol.27, pp.379, 623, (1948).

C. Shannon "Communications in the presence of noise", *Proc. IRE*, vol. 37, pp.10-21 (1949).

D. Slepian "On bandwidth", *Proc. IEEE*, vol. 64, pp.292-300 (1976).

E.T. Whittaker "On the functions which are represented by the expansions of the interpolation theory", *Proc. Royal Society of Edinburg*, vol. 35, pp.181-194 (1915).

J.M. Whittaker "The Fourier theory of the cardinal functions", *Proc. the Math. Soc. of Edinburg*, vol. 1, pp.169-176 (1929).

J.M. Whittaker, **Interpolatory Function Theory**, Cambridge University Press (Cambridge Tracts in Mathematics and Mathematical Physics), Cambridge, no.33, 1935.

2

Fundamentals of Fourier Analysis and Stochastic Processes

The sampling theorem is traditionally expressed in the language of Fourier analysis. Stochastic processes are used to analyze the effects of uncertainty on the interpolation process and, in certain cases, are used as signal models. This chapter contains a brief overview of the fundamentals of Fourier analysis and stochastic processes. The material in this chapter is prerequisite for material in the remainder of the book. Except for an isolated application in Section 3.5, however, stochastic processes are not used until Chapter 5.

2.1 Signal Classes

There are a number of signal classes to which we will make common reference:

(a) **Periodic Signals**

A signal, $x(t)$, is periodic if there exists a T such that

$$x(t) = x(t - T)$$

for all t. The function $x(t) = constant$ is periodic.

(b) **Finite Energy Signals**[1]

If

$$E = \int_{-\infty}^{\infty} |x(t)|^2 \, dt < \infty,$$

then $x(t)$ is said to have finite energy E.

(c) **Finite Area Signals**[1]

If,

$$A = \int_{-\infty}^{\infty} |x(t)| \, dt < \infty,$$

then $x(t)$ is said to have finite area A.

(d) **Bounded Signals**[1]

If, for C a constant,

$$|x(t)| \leq C < \infty,$$

then $x(t)$ is said to be bounded.

(e) **Bandlimited Signals**

If there exists a finite (bandwidth) B such that

$$\int_{-\infty}^{\infty} x(t)e^{-j2\pi ut} \, dt = 0 \; ; \; |u| > B,$$

then $x(t)$ is said to be bandlimited in the low pass sense.

(f) **Analytic Signals**

If, for any finite complex number φ, we have equality in the Taylor series

$$x(z) = \sum_{n=0}^{\infty} \frac{x^{(n)}(\varphi)\,(z - \varphi)^n}{n!}$$

where

$$x^{(n)}(z) = \left(\frac{d}{dz}\right)^n x(z)$$

---

[1]The classes of finite energy, finite area and bounded signals (when Lebesgue measurable) are recognized respectively as L_2, L_1 and L_∞ signals [Naylor and Sell; Luenberger].

then $x(z)$ is said to be analytic everywhere (in the closed z plane).

Except for bandlimited functions being subsumed in the class of analytic functions, membership in one of these six classes does not necessarily dictate membership in another. For example, $\log(t)$ on the interval $(0, 1)$ has finite area and energy and yet it is not bounded. Other examples are given as exercises. Except for the degenerate case of $x(t) \equiv 0$, a periodic signal can have neither finite energy nor finite area.

2.2 The Fourier Transform

A Fourier transform can reveal if a signal can be represented uniquely in terms of its samples. In this chapter, we review the basic properties of Fourier transforms for the purposes of establishing notation and completeness. More in depth treatments of the topic can be found in the classic books by Bracewell and Papoulis (1962) or in any one of the numerous texts on the topic [Papoulis (1980), Liu and Liu, Oppenheim and Willsky].

The Fourier transform (or spectrum) of a signal $x(t)$ is

$$X(u) = \int_{-\infty}^{\infty} x(t)\, e^{-j2\pi ut}\, dt. \qquad (2.1)$$

Fourier transform pairs will be denoted by

$$x(t) \longleftrightarrow X(u).$$

The inversion formula is

$$x(t) = \int_{-\infty}^{\infty} X(u)\, e^{j2\pi ut}\, du.$$

From the transform definitions, the Fourier transform theorems in Table 2.1 can be generated. Proofs are left as an exercise.

	$x(t)$	$\longleftrightarrow$	$X(u)$
transform	$x(t)$	$\longleftrightarrow$	$\int\limits_{-\infty}^{\infty} x(t)e^{-j2\pi ut}dt$
scaling	$x(at)$	$\longleftrightarrow$	$\frac{1}{|a|}X(\frac{u}{a})$
shift	$x(t-\tau)$	$\longleftrightarrow$	$X(u)e^{-j2\pi u\tau}$
derivative	$\left(\frac{d}{dt}\right)^n x(t)$	$\longleftrightarrow$	$(j2\pi u)^n X(u)$
integral	$\int\limits_{-\infty}^{t} x(\tau)d\tau$	$\longleftrightarrow$	$\frac{X(u)}{j2\pi u} + \frac{1}{2}X(0)\delta(u)$
conjugate	$x^*(t)$	$\longleftrightarrow$	$X^*(-u)$
transpose	$x(-t)$	$\longleftrightarrow$	$X(-u)$
convolution	$x(t) * h(t)$	$\longleftrightarrow$	$X(u)H(u)$
inversion	$\int_{-\infty}^{\infty} X(u)e^{j2\pi ut}du$	$\longleftrightarrow$	$X(u)$
duality	$X(t)$	$\longleftrightarrow$	$x(-u)$
linearity	$ax_1(t) + bx_2(t)$	$\longleftrightarrow$	$aX_1(u) + bX_2(u)$
correlation	$x(t) \star h(t)$	$\longleftrightarrow$	$X(u)H^*(u)$

Table 2.1: Fourier transform theorems. Convolution is defined by

$$x(t) * h(t) = \int_{-\infty}^{\infty} x(\tau) h(t-\tau)\, d\tau.$$

Correlation is

$$x(t) \star h(t) = \int_{-\infty}^{\infty} x(\tau)h^*(\tau - t)d\tau$$
$$= x(t) * h^*(-t).$$

2.2.1   The Fourier Series

Periodic functions with period T can be expressed as a Fourier series,

$$x(t) = \sum_{n=-\infty}^{\infty} c_n\, e^{j2\pi nt/T}, \tag{2.2}$$

where

$$c_n = \frac{1}{T} \int_T x(t)\, e^{-j2\pi nt/T}\, dt, \tag{2.3}$$

and integration is over any single period.

If we define $x_T(t)$ to be a single period of $x(t)$:

$$x_T(t) = \begin{cases} x(t) & ; \quad \tau < t \le \tau + T \\ 0 & ; \quad \text{otherwise} \end{cases}$$

(τ is arbitrary), then (2.3) can be written

$$\begin{aligned} c_n &= \frac{1}{T} \int_{-\infty}^{\infty} x_T(t)\, e^{-j2\pi nt/T}\, dt \\ &= \frac{1}{T} X_T\!\left(\frac{n}{T}\right). \end{aligned} \tag{2.4}$$

The Fourier coefficients can thus be determined by sampling the Fourier transform of any period of the periodic function.

2.2.1.1   Convergence

If on any interval the length of a period, $x(t)$ has finite energy, then (2.2) is better written as

$$\lim_{N \to \infty} \int_T |x(t) - x_N(t)|^2\, dt = 0 \tag{2.5}$$

where the partial sum is

$$x_N(t) = \sum_{n=-N}^{N} c_n\, e^{j2\pi nt/T}$$

That is, convergence is in the mean square sense.

2.2.1.2 Orthogonal Basis Functions

The Fourier series is a special case of an orthogonal basis function expansion [Luenberger; Naylor and Sell]. For a given interval, I, functions $x(t)$ and $y(t)$ are said to be *orthogonal* if

$$\int_I x(t)\, y^*(t)dt = 0$$

where the superscript asterisk denotes complex conjugation. The functions are said to be *orthonormal* if both $x(t)$ and $y(t)$ have unit energy on the interval I.

Each element in an orthonormal basis set, $\{\varphi_n(t)|-\infty < n < \infty; t \in I\}$ [2], has unit energy on the interval I and is orthogonal to every other element in the set:

$$\int_I \varphi_n(t)\, \varphi_m^*(t)dt = \delta[n - m], \tag{2.6}$$

where the *Kronecker delta* is

$$\delta[n] = \left\{ \begin{array}{ll} 1 & ;\quad n = 0 \\ 0 & ;\quad n \neq 0. \end{array} \right.$$

Corresponding to a given class, C, of finite energy functions, a basis set is said to be *complete* if, for every $x(t) \in C$, we can write

$$x(t) = \sum_{n=-\infty}^{\infty} \alpha_n\, \varphi_n(t) \; ;\; t \in I \tag{2.7}$$

where equality is at least in the mean square sense:

$$\lim_{N \to \infty} \int_I |x(t) - \sum_{n=-N}^{N} \alpha_n\varphi_n(t)|^2 \, dt = 0. \tag{2.8}$$

The expansion coefficients, α_n, can be found by multiplying both sides of (2.7) by $\varphi_m^*(t)$ and integrating over I. Using (2.6), we find that

$$\alpha_n = \int_I x(t)\varphi_n^*(t) \, dt.$$

Examples: In each of the following examples, C is subsumed in the class of finite energy (L_2) functions.

---

[2]For certain orthogonal basis sets such as the prolate spheroidal wave functions in Chapter 7, the index n runs from 0 to ∞.

- The cardinal series is an orthonormal expansion for all signals whose Fourier transforms are bandlimited with bandwidth B. For $I = \{t | -\infty < t < \infty\}$, the basis functions are:

$$\varphi_n(t) = \frac{1}{\sqrt{2B}} \frac{\sin[\pi(2Bt - n)]}{\pi(2Bt - n)}.$$

Note, then, that $\alpha_n = \frac{1}{\sqrt{2B}} x(\frac{n}{2B})$. As will be shown in Section (3.3.1), the cardinal series displays uniform convergence which is stronger than that in (2.8).

- The Fourier series is an orthonormal expansion for signals over the interval $I = \{t | -\frac{T}{2} < t < \frac{T}{2}\}$. The basis function here are:

$$\varphi_n(t) = \frac{1}{\sqrt{T}} exp\left(\frac{j2\pi nt}{T}\right) ; t \in I.$$

It follows that $\alpha_n = \frac{1}{\sqrt{T}} X_T(\frac{n}{T})$.

The *prolate spheroidal wave functions* in section 7.2 and the complex Walsh functions in Exercise 4.23 are also examples of orthogonal basis sets. Such sets will later prove useful in the understanding of interpolation from continuous samples and Kramer's generalization of the sampling theorem.

2.2.2   Some Elementary Functions

In this section we define a number of elementary functions. Each is pictured in Fig. 2.1 . Their Fourier transforms will be developed in the next section.

(a) **The Rectangle Function**

$$\Pi(t) = \begin{cases} 1 & ; \ |t| < \frac{1}{2} \\ 0 & ; \ |t| > \frac{1}{2} \\ \frac{1}{2} & ; \ |t| = \frac{1}{2}. \end{cases}$$

(b) **The Sinc Function**

$$\text{sinc}(t) = \frac{\sin(\pi t)}{\pi t}.$$

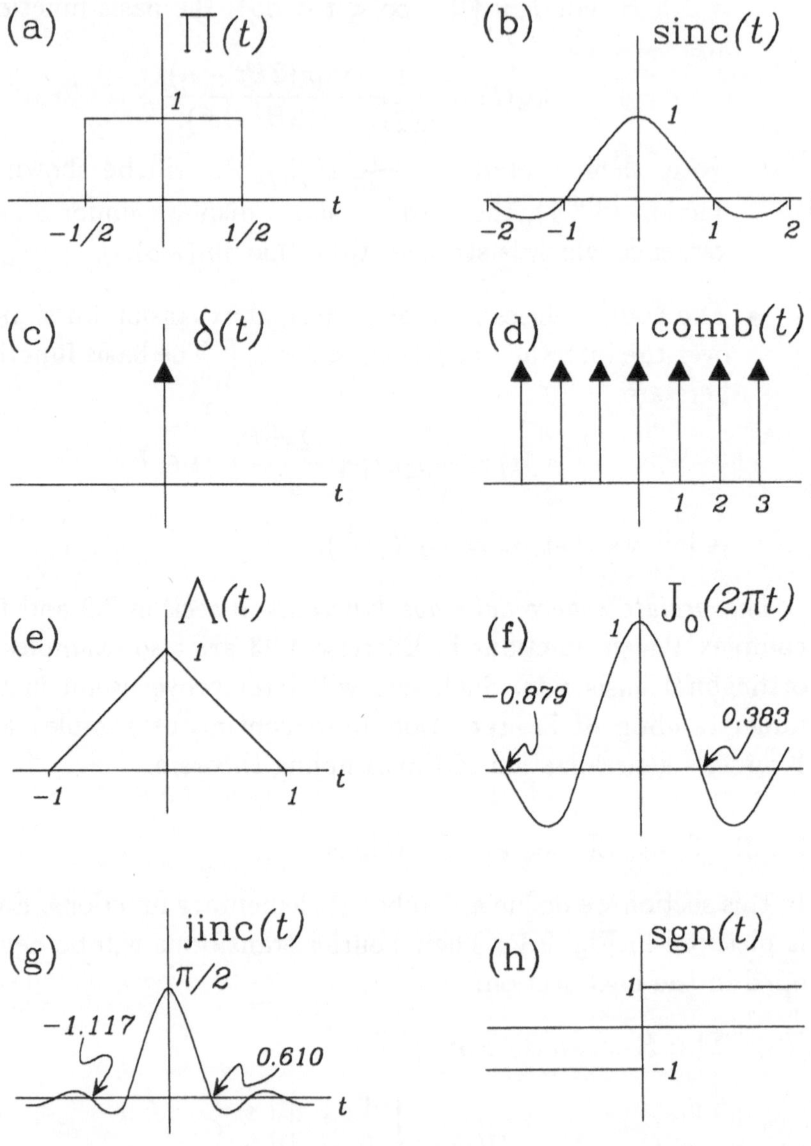

Figure 2.1: Some elementary functions.

One advantage of this notation is that, for n an integer,

$$\text{sinc}(n) = \delta[n]. \qquad (2.9)$$

(c) The Dirac Delta Function

Rigorously, the Dirac delta, $\delta(t)$, is a distribution and not a function. It can be defined by its sifting property. If $x(t)$ is continuous at $\tau = t$, then

$$x(t) = \int_{-\infty}^{\infty} x(\tau)\delta(t - \tau)\,d\tau. \qquad (2.10)$$

Note we use parenthesis for the Dirac delta argument and square brackets for the Kronecker delta.

The area of a Dirac delta can be found from the sifting property by setting $x(\tau) = 1$ and $t = 0$. Recognizing further that $\delta(t) = 0$ except at the origin leads us to conclude that for any $\varepsilon > 0$

$$\int_{-\varepsilon}^{\varepsilon} \delta(\tau)\,d\tau = 1. \qquad (2.11)$$

The Dirac delta can be viewed as the limit of any one of a number of unit area functions which approach zero width and infinite height at the origin. For example

$$\delta(t) = \lim_{A\to\infty} A\ \Pi(At)$$

or, alternately

$$\delta(t) = \lim_{A\to\infty} A\ \text{sinc}(At).$$

Using the definition in (2.11), it is easy to show that

$$\delta(t) = |a|\ \delta(at).$$

Lastly, since

$$\exp(j2\pi\xi t) \longleftrightarrow \delta(u - \xi), \qquad (2.12)$$

we conclude from (2.2) that the spectrum of a periodic signal can be written as a string of weighted Dirac deltas:

$$X(u) = \sum_{n=-\infty}^{\infty} c_n \, \delta\left(u - \frac{n}{T}\right).$$

(d) The Comb Function

$$\text{comb}(t) = \sum_{n=-\infty}^{\infty} \delta(t - n).$$

Any periodic function can be written as

$$x(t) = x_T(t) * \frac{1}{T} \, \text{comb}\left(\frac{t}{T}\right). \tag{2.13}$$

(e) The Triangle Function

$$\Lambda(t) = (1 - |t|) \, \Pi\left(\frac{t}{2}\right).$$

Note that

$$\Lambda(t) = \Pi(t) * \Pi(t). \tag{2.14}$$

(f) Bessel Functions

For $\nu > -\frac{1}{2}$, Bessel functions of the first kind can be defined by the integral [Abramowitz & Stegun]

$$J_\nu(z) = \frac{2(\frac{z}{2})^\nu}{\sqrt{\pi} \, \Gamma(\nu + \frac{1}{2})} \int_0^1 (1 - u^2)^{\nu - \frac{1}{2}} \cos(zu) \, du \tag{2.15}$$

where *the gamma function* for positive argument is defined as

$$\Gamma(\xi) = \int_0^\infty \tau^{\xi-1} e^{-\tau} \, d\tau.$$

When $\xi = n$ is a non-negative integer, $\Gamma(n + 1) = n!$. When $\xi = n + \frac{1}{2}$,

$$\Gamma\left(n + \frac{1}{2}\right) = \frac{(2n - 1)!! \, \sqrt{\pi}}{2^n} \tag{2.16}$$

where, for example, $7!! = 7 \cdot 5 \cdot 3 \cdot 1$. Equation (2.16) can be derived from the property $\Gamma(\xi + 1) = \xi\Gamma(\xi)$ and $\Gamma(\frac{1}{2}) = \sqrt{\pi}$.

For $\nu = 0$, (2.15) becomes

$$J_0(2\pi t) = \frac{2}{\pi} \int_0^1 \frac{\cos(2\pi ut)}{\sqrt{1 - u^2}} \, du \qquad (2.17)$$

which is shown in Fig. 2.1f.

(g) The Jinc Function

We define [Bracewell]

$$\text{jinc}(t) = \frac{J_1(2\pi t)}{2t}. \qquad (2.18)$$

Using (2.15) with $\nu = 1$ gives

$$\text{jinc}(t) = 2 \int_0^1 \sqrt{1 - u^2} \, \cos(2\pi ut) \, du.$$

Since

$$2 \int \sqrt{1 - u^2} \, du = u\sqrt{1 - u^2} + \arcsin(u), \qquad (2.19)$$

it follows that

$$\text{jinc}(0) = \frac{\pi}{2}.$$

The zero crossings of the jinc are identical to the well tabulated $t \neq 0$ zero crossings of $J_1(2\pi t)$.

(h) The Signum Function

The function

$$\text{sgn}(t) = \begin{cases} -1 & ; \quad t < 0 \\ 0 & ; \quad t = 0 \\ 1 & ; \quad t > 0 \end{cases}$$

is denoted by a contraction of the word *sign* and is pronounced *signum* to avoid confusion with the trigonometric *sin*.

Note that the *unit step function* can be written

$$\mu(t) = \frac{1}{2}[\text{sgn}(t) + 1]. \qquad (2.20)$$

2.2.3   Some Transforms of Elementary Functions

A list of Fourier transform pairs is in Table 2.2. Some can easily be generated from the function and Fourier transform definitions. We will elaborate only on three of the more challenging entries. Others will be left as exercises.

More extensive lists of transform pairs exist elsewhere [Bracewell]. A wealth of Fourier transform pairs can be obtained from *Laplace transform* tables [Abramowitz and Stegun]. The (unilateral) Laplace transform is defined by

$$X_L(s) = \int_0^\infty x(t)e^{-st}dt \; ; \quad s = \sigma + j2\pi u. \tag{2.21}$$

If

(1) $x(t)$ is *causal* [i.e. $x(t) = x(t)\,\mu(t)$] and

(2) $X_L(s)$ converges for $\sigma = 0$,

then the Laplace transform evaluated at $\sigma = 0$ is the Fourier transform of $x(t)$. That is,

$$X_L(j2\pi u) = X(u).$$

A sufficient condition for the second criterion to hold is that $x(t)$ have finite area:

$$A = \int_{-\infty}^\infty |x(t)| \, dt < \infty.$$

This follows from the inequality

$$\left| \int_{-\infty}^\infty y(t)dt \right| \le \int_{-\infty}^\infty |y(t)|dt$$

which, when applied to (2.21) yields

$$|X_L(j2\pi u)| = \left| \int_{-\infty}^\infty x(t)e^{-j2\pi ut}dt \right| \le \int_{-\infty}^\infty |x(t)|dt < \infty.$$

$x(t) \longleftrightarrow X(u)$	
$\delta(t)$	1
$\Pi(t)$	$\text{sinc}(u)$
$\Lambda(t)$	$\text{sinc}^2(u)$
$\text{sgn}(t)$	$-\frac{j}{\pi u}$
$\mu(t)$	$\frac{1}{2}[\delta(u) - \frac{j}{\pi u}]$
$\text{jinc}(t)$	$\sqrt{1 - u^2}\,\Pi\left(\frac{u}{2}\right)$
$\exp(-\pi t^2)$	$\exp(-\pi u^2)$
$J_0(2\pi t)$	$\frac{1}{\pi\sqrt{1-u^2}}\Pi(\frac{u}{2})$
$\exp(-|t|)$	$2[1 + (2\pi u)^2]^{-1}$
$\text{comb}(t)$	$\text{comb}(u)$

Table 2.2: Some Fourier transform pairs

(a) The Signum Function

For sgn(t), the Fourier transform becomes

$$X(u) = -j \int_{-\infty}^{\infty} \text{sgn}(t) \sin(2\pi ut) dt$$

where we have expanded the exponential via Euler's formula and have recognized the odd component of the integrand integrates to zero. The evenness of the remaining integrand above can be exploited to write:

$$
\begin{aligned}
X(u) &= -j2 \int_{0}^{\infty} \sin(2\pi ut)\, dt \\
&= j\Im[2 \int_{0}^{\infty} e^{-j2\pi ut}\, dt] \\
&= \lim_{\alpha \to 0} j\Im[2 \int_{0}^{\infty} e^{-(\alpha+j2\pi u)t}\, dt] \\
&= \lim_{\alpha \to 0} j\Im[\frac{2}{\alpha + j2\pi u}] \\
&= \frac{1}{j\pi u}
\end{aligned}
$$

where $\Im$ denotes the imaginary component operation.

(b) The Gaussian

Differentiating the expression for the Fourier transform of the Gaussian, $x(t) = exp(-\pi t^2)$, gives

$$
\begin{aligned}
\frac{dX(u)}{du} &= \int_{-\infty}^{\infty} (-j2\pi t) \exp[-\pi(t^2 + j2ut)]\, dt \\
&= -j \int_{-\infty}^{\infty} \pi(2t + j2u) \exp[-\pi(t^2 + j2ut)]dt - 2\pi u X(u).
\end{aligned}
$$

The resulting integral can be evaluated in closed form.

$$\frac{dX(u)}{du} = j \exp[-\pi(t^2 + j2ut)]|_{-\infty}^{\infty} - 2\pi u X(u) = -2\pi u X(u)$$

or

$$\frac{dX(u)}{X(u)} = -2\pi u\, du.$$

Integrating both sides gives

$$\ln[X(u)] = -\pi u^2$$

from which the corresponding entry in Table 2.2 follows.

(c) **The Comb Function**

Since $\mathrm{comb}(t)$ is periodic, it can be expressed in terms of a Fourier series. The coefficients are $c_n = 1$ for all n. Thus

$$\mathrm{comb}(t) = \sum_{n=-\infty}^{\infty} e^{j2\pi nt}. \qquad (2.22)$$

Using (2.12), it follows that

$$\mathrm{comb}(t) \longleftrightarrow \sum_{n=-\infty}^{\infty} \delta(u-n) = \mathrm{comb}(u).$$

2.2.4   Other Properties

(a) The **Power Theorem** is

$$\int_{-\infty}^{\infty} x(t)y^*(t)\,dt = \int_{-\infty}^{\infty} X(u)Y^*(u)\,du.$$

An important special case is *Parseval's theorem* :

$$\int_{-\infty}^{\infty} |x(t)|^2 dt = \int_{-\infty}^{\infty} |X(u)|^2 du.$$

For a proof of the power theorem, we write

$$\int_{-\infty}^{\infty} x(t)y^*(t)dt = \int_{-\infty}^{\infty} \left[\int_{-\infty}^{\infty} X(u)e^{j2\pi ut}du\right] y^*(t)\,dt.$$

Reversing integration order and recognizing the resulting transform completes the proof.

(b) The **Poisson Sum Formula** is

$$T\sum_{n=-\infty}^{\infty} x(t-nT) = \sum_{n=-\infty}^{\infty} X\left(\frac{n}{T}\right) e^{j2\pi nt/T}. \qquad (2.23)$$

A proof follows:

$$\sum_{n=-\infty}^{\infty} x(t - nT) = x(t) * \frac{1}{T}\text{comb}\left(\frac{t}{T}\right)$$

$$= x(t) * \frac{1}{T}\sum_{n=-\infty}^{\infty} e^{j2\pi nt/T}$$

where we have used (2.22). Recognizing

$$x(t) * e^{j2\pi\nu t} = X(\nu)e^{j2\pi\nu t}$$

completes the proof.

Note that, for $t = 0$, the Poisson sum formula directly relates the sum of the signal and spectral samples:

$$T \sum_{n=-\infty}^{\infty} x(nT) = \sum_{n=-\infty}^{\infty} X\left(\frac{n}{T}\right). \qquad (2.24)$$

2.3   Stochastic Processes

In this section, we will briefly review the stochastic process and its characterization. More in depth treatment is available in numerous excellent texts [Papoulis (1984); Parzen; Thomas; Stark and Woods].

A stochastic process can be either discrete or continuous and can be used to model either noise or a signal.

2.3.1   First and Second Order Statistics

In many applications, the first and second order statistics of a stochastic process suffice for a description. The mean of a continuous time stochastic process, $\xi(t)$, is

$$\overline{\xi(t)} = E[\xi(t)]$$

where E denotes the expected value operator. The *noise level* is the variance of the process:

$$\text{var }\xi(t) = E[|\xi(t) - \overline{\xi(t)}|^2]$$

$$= \overline{|\xi(t)|^2} - |\overline{\xi(t)}|^2.$$

If a process is *zero mean*, then $\overline{\xi(t)}$ is zero and the noise level is $\overline{|\xi(t)|^2}$. If a process is a signal, this is called the *signal level*.

Both the mean and noise (signal) level are first order statistics since they are only concerned about the process at a single point in time. Second order statistics are concerned about the process at two different points in time. The *autocorrelation*, for example, is

$$R_\xi(t;\tau) = E[\xi(t)\xi^*(\tau)].$$

Note that

$$R_\xi(t;t) = \overline{|\xi(t)|^2}. \tag{2.25}$$

Similarly, the cross correlation between two processes, $\xi(t)$ and $\eta(t)$, is defined as

$$R_{\xi\eta}(t;\tau) = E[\xi(t)\eta^*(\tau)].$$

2.3.2   Stationary Processes

A process that does not change its character with respect to time is said to be *stationary*. A process is said to be *stationary in the wide sense* if it meets two conditions. First, the mean of the process must be a constant for all time:

$$\overline{\xi(t)} = \overline{\xi}.$$

Secondly, the autocorrelation is only a function of the distance between the two points of interest:

$$E[\xi(t)\xi^*(\tau)] = R_\xi(t - \tau). \tag{2.26}$$

For such processes, the autocorrelation can thus be described by a one dimensional rather than a two dimensional function.

2.3.2.1   Power Spectral Density

The *power spectral density*, $S_\xi(u)$, of a wide sense stationary process can be defined as the Fourier transform of the autocorrelation.

$$R_\xi(t) \longleftrightarrow S_\xi(u).$$

The power spectral density measures the power content of the process at each frequency component.

One can easily show that the autocorrelation is *hermetian*:

$$R_\xi(t) = R_\xi^*(-t). \tag{2.27}$$

It follows immediately that $S_\xi(u)$ is real. It is also non-negative.

For a wide sense stationary discrete process, the autocorrelation is

$$R_\xi[n - m] = E\left\{\xi[n]\,\xi^*[m]\right\}.$$

The power spectral density for such processes is given by a Fourier series:

$$S_\xi(u) = \sum_{n=-\infty}^{\infty} R_\xi[n]e^{-j2\pi n u}. \tag{2.28}$$

For both discrete and continuous wide sense stationary processes, the noise level can be obtained by setting the argument of the autocorrelation to zero if the process is zero mean.

2.3.2.2 Some Stationary Noise Models

Here we list some autocorrelation functions for later use. In general, one can choose any non-negative finite energy function for $S_\xi(u)$. If the process is zero mean, the power spectrum's inverse transform is a valid autocorrelation function with

$$\overline{|\xi(t)|^2} = \int_{-\infty}^{\infty} S_\xi(u)\,du. \tag{2.29}$$

(a) **Stationary White Noise**

Stationary white noise has an autocorrelation of

$$R_\xi(\tau) = \overline{|\xi|^2}\,\delta(\tau). \tag{2.30}$$

Only for white noise does the notation $\overline{|\xi|^2}$ <u>not</u> correspond to the second moment of the process. Indeed, the noise level for this process is infinite. White noise is so named because its power spectral density

$$S(u) = \overline{|\xi|^2} \tag{2.31}$$

has the same energy level at every frequency.

(b) Stationary Discrete White Noise

This is a discrete process with autocorrelation

$$R_\xi[n] = \overline{|\xi|^2}\, \delta[n].$$ (2.32)

Note that one does not obtain a discrete white sequence by sampling a continuous white noise process. Here, for example, the noise level is finite.

(c) Laplace Autocorrelation

For a given positive parameter λ, the Laplace autocorrelation is

$$R_\xi(\tau) = \overline{|\xi|^2}\, e^{-\lambda|\tau|}.$$ (2.33)

It follows that

$$S_\xi(u) = \frac{2\,\lambda\,\overline{|\xi|^2}}{\lambda^2 + (2\pi u)^2}.$$ (2.34)

2.3.2.3 Linear Systems with Stationary Stochastic Inputs

Let $h(t)$ be a deterministic signal and define

$$\eta(t) = \xi(t) * h(t).$$ (2.35)

Then the cross correlation between input and output is

$$R_{\eta\xi}(\tau) = R_\xi(\tau) * h(\tau)$$ (2.36)

or, in terms of the power spectral densities:

$$S_{\eta\xi}(u) = S_\xi(u)H(u).$$

The output's autocorrelation is

$$R_\eta(t) = [h(t) \star h(t)] * R_\xi(t)$$ (2.37)

or, in the frequency domain,

$$S_\eta(u) = |H(u)|^2\, S_\xi(u).$$ (2.38)

2.4   Exercises

2.1 Evaluate $\int_{-\infty}^{\infty} x(t)dt$ for each of the functions in Fig. 2.1 that have finite energy.

2.2 Define $\delta(t)$ as the limit of a

   (a) triangle function.

   (b) jinc.

   (c) sinc squared.

2.3 Derive the entries in Table 2.1.

2.4 Derive the entries in Table 2.2 for

   (a) $\delta(t)$.

   (b) $\Pi(t)$.

   (c) $\Lambda(t)$.

   (d) $\mu(t)$.

   (e) $jinc(t)$.

2.5 Assume $g(t)$ has a single zero crossing at $t = t_0$, i.e. $g(t_0) = 0$. Let $\frac{dg(t_0)}{dt} = a$.

   (a) Simplify the expression for $\delta(g(t))$.

   (b) Evaluate $\delta(\ln(t/b))$ for both t and b positive.

   (c) Evaluate $\pi\delta(\sin(\pi t))$.

2.6 Evaluate the transforms of

   (a) $sinc(t)$.

   (b) $x(\frac{t-a}{b})$.

2.7 (a) Find the area of

   i. $sinc^4(t)$.

   ii. $jinc^2(t)$.

   iii. $J_0^2(2\pi t)$.

(b) Find the integral of sinc(t)jinc(t) over all time.

2.8 Evaluate the series $\sum\limits_{n=-\infty}^{\infty} a_n$ when $a_n =$

(a) $\text{sinc}(t - n)$.

(b) $\text{sinc}^2(t - n)$.

(c) $\text{jinc}(t - n)$.

(d) $\text{sinc}[a(t - n)]$.

(e) $\text{jinc}(2n)$.

(f) $\text{jinc}^2(\frac{n}{4})$.

(g) $n \exp[-\pi(nT)^2]$.

2.9 (a) Find the first few extrema of $\text{sinc}(t)$.

(b) For large t, find a good approximation for the extrema of $\text{sinc}(t)$.

2.10 At a finite discontinuity, a Fourier series converges to the arithmetic midpoint. To illustrate, let

$$y(t) = \sum_{n=-\infty}^{\infty} \Pi\left(\frac{t - nT}{\tau}\right).$$

Express $y(t)$ as a Fourier series and evaluate $y(\frac{\tau}{2})$ using that series. Let $\alpha = \tau/T < 1$.

2.11 Evaluate the Fourier transform of

$$\frac{J_\nu(2\pi t)}{(2t)^\nu} \; ; \; \nu > -\frac{1}{2}.$$

Simplify your result for $\nu = 0$ and $\nu = 1$ thus deriving two of the entries in Table 2.2.

2.12 *Gibb's Phenomena*: From Exercise 2.10, let $\alpha = \frac{1}{2}$ and

$$z(t) = 2y\left(t - \frac{\tau}{2}\right) - 1.$$

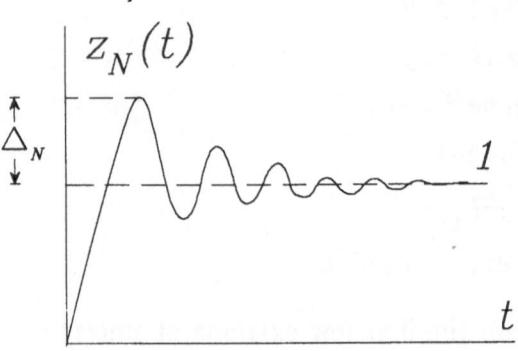

Figure 2.2: Gibb's Phenomena.

The truncated Fourier series for $z(t)$ is $z_N(t)$ which will display overshoot, Δ_N, as shown in Figure 2.2. Compute this overshoot and its exact value as $N \to \infty$.

HINT: $Si(\pi) = 1.851937$ where

$$Si(z) = \int_0^z \frac{\sin(t)}{t} dt.$$

2.13 (a) Derive (2.29).

(b) Derive a similar expression for a discrete stochastic process.

2.14 Prove (2.36) and (2.37).

2.15 Show that
$$|R_\xi(\tau)| \le \overline{\xi^2}.$$

2.16 (a) Show that (2.27) is valid.

(b) Show that, as a consequence, $S_\xi(u)$ is real.

2.17 Show that

$$\text{comb}\left(t - \frac{1}{2}\right) \leftrightarrow \text{comb}\left(\frac{u}{2}\right) - \text{comb}\left(\frac{u-1}{2}\right).$$

2.18 (a) Let $x(t) = t^{-\frac{3}{4}}$. Compute both the area and energy of $x(t)$

     i . over the interval $(0,1)$ and,

    ii . over the interval $(1, \infty)$.

(b) Use your result in (a) to show that finite area does not imply finite energy and visa versa.

(c) The energy of a sequence of samples can be defined as in (3.32). Let

$$A = \frac{1}{2B} \sum_{n=-\infty}^{\infty} \left| x\left(\frac{n}{2B}\right) \right|.$$

If $A < \infty$, can we conclude that $E < \infty$? What about the converse?

2.19 (a) If $x(t)$ has a spectrum with finite area, show that $x(t)$ is bounded. Let

$$|x(t)| \leq C < \infty.$$

Express C in terms of

$$A = \int_{-\infty}^{\infty} |X(u)| du < \infty.$$

(b) Show that the converse of (a) is not true. That is, there exist bounded signals whose spectra do not have finite area.

2.20 Show that a finite energy bandlimited signal, $x(t)$, must be bounded. Specifically,

$$|x(t)|^2 \leq 2BE.$$

HINT: Apply Schwarz's inequality to the inversion formula.

2.21 (a) A finite energy signal $x(t)$ is bandlimited. Is its pth derivative also a finite energy bandlimited function?

   (b) Repeat part (a) substituting the word "area" for "energy".

   (c) Substitute "bounded".

2.22 A series

$$\sum_{m=0}^{\infty} a_m$$

is said to be *absolutely convergent* if the sum

$$S = \sum_{m=0}^{\infty} |a_m|$$

is finite. The terms in an absolutely convergent series can be arbitrarily rearranged without affecting the sum. The Taylor series expansion of $x(t)$ about the (real) number τ is

$$x(t) = \sum_{m=0}^{\infty} \frac{(t-\tau)^m}{m!} x^{(m)}(\tau).$$

Show that if $x(t)$ has finite energy and is bandlimited, then this series is absolutely convergent for all $|t| < \infty$.

HINT: Show that

$$S(t) < \sqrt{2B} E e^{2\pi B|t-\tau|}.$$

2.23 (a) Apply Schwarz's inequality to the derivative theorem to show that, if $x(t)$ is bandlimited, its Mth derivative is bounded as

$$|x^{(M)}(t)|^2 \leq \frac{(2\pi B)^{2M+1} E}{2M + 1}. \qquad (2.39)$$

   (b) Show that a bandlimited function is smooth in the sense that

$$|x(t+\tau) - x(t)|^2 \leq \frac{(2\pi B)^3 |\tau|^2 E}{3}. \qquad (2.40)$$

2.24 Consider the general linear integral transform

$$g(t) = \int_{-\infty}^{\infty} f(\tau)h(t;\tau)d\tau.$$

For a given t, assume that the kernel, $h(t;\tau)$, as a function of τ, has finite energy,

$$E_h(t) = \int_{-\infty}^{\infty} |h(t;\tau)|^2 d\tau.$$

If $f(t)$ has finite energy, show that $g(t)$ is bounded.

REFERENCES

M. Abramowitz and I.A. Stegun, **Handbook of Mathematical Functions**, Dover, New York, 9th.ed., 1964.

R.N. Bracewell, **The Fourier Transform and Its Applications**, 2nd edition, Revised, McGraw-Hill, NY, 1986.

C. L. Liu and Jane W. S. Liu, **Linear Systems Analysis**, McGraw-Hill, New York, 1975.

D.G. Luenberger, **Optimization by Vector Space Methods** Wiley, New York, 1969.

A.W. Naylor and G.R. Sell, **Linear Operator Theory in Engineering and Science**, Springer-Verlag, New York, 1982.

A. V. Oppenheim, Alan S. Willsky and Ian T. Young, **Signals and Systems**, Prentice-Hall, New Jersey, 1983.

A. Papoulis, **The Fourier Integral and Its Applications**, McGraw-Hill, New York, 1962.

A. Papoulis, **Circuits and Systems, A Modern Approach**, Holt, Rinehart and Winston, Inc., 1980.

A. Papoulis, **Probability, Random Variables, and Stochastic Processes**, 2nd Edition, McGraw-Hill, New York, 1984.

E. Parzen, **Stochastic Processes**, Holden-Day, San Francisco, 1962.

H. Stark and John W. Woods, **Probability, Random Processes, and Estimation Theory For Engineers**, Prentice-Hall, New Jersey, 1986.

J.B. Thomas, **An Introduction to Statistical Communication Theory**, John Wiley & Sons, Inc., NY, 1969.

3

The Cardinal Series

A signal is bandlimited in the low pass sense if there is a $B > 0$ such that

$$X(u) = X(u) \, \Pi\left(\frac{u}{2B}\right). \tag{3.1}$$

That is, the spectrum is identically zero for $|u| > B$. The B parameter is referred to as the signal's *bandwidth*. It then follows that

$$x(t) = \int_{-B}^{B} X(u) \, e^{j2\pi ut} \, du. \tag{3.2}$$

In most cases, the signal can be expressed by the cardinal series

$$x(t) = \sum_{n=-\infty}^{\infty} x\left(\frac{n}{2B}\right) \operatorname{sinc}(2Bt - n). \tag{3.3}$$

The ability to thus express a continuous signal in terms of its samples is the fundamental statement of the sampling theorem.

At other than sample locations, a more computationally efficient form of (3.3) requiring evaluation of only a single trigonometric function is

$$x(t) = \frac{1}{\pi} \sin(2\pi Bt) \sum_{n=-\infty}^{\infty} \frac{(-1)^n \, x\left(\frac{n}{2B}\right)}{2Bt - n}. \tag{3.4}$$

3.1 Interpretation

The sampling theorem reduces the normally continuum infinity of ordered pairs required to specify a function to a countable – although still infinite – set. Remarkably, these elements are obtained directly by sampling.

Bandlimited functions are smooth. Any behavior deviating from "smooth" would result in high frequency components which

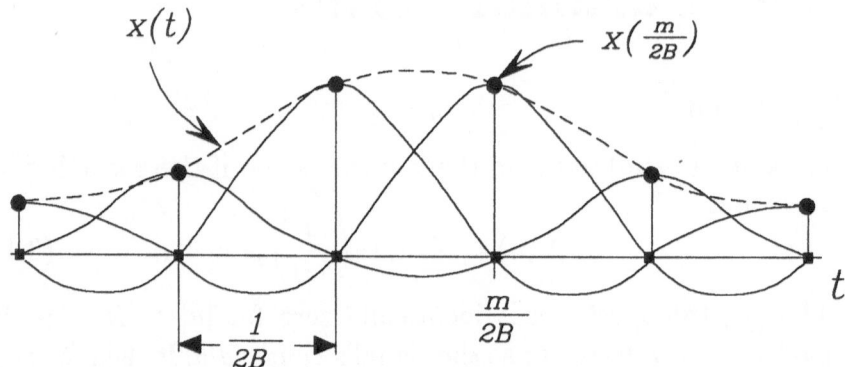

Figure 3.1: Illustration of the manner by which $x(t)$ is regained from its samples using the cardinal series. Note that the sinc for a given sample is zero at all other sample locations (shown here as squares).

in turn invalidate the required property of being bandlimited. The smoothness of the signal between samples precludes arbitrary variation of the signal there.

Let's examine the cardinal series more closely. Evaluation of (3.3) at $t = m/2B$ and using (2.9) reduces (3.3) to an identity. Thus, only the sample at $t = m/2B$ contributes to the interpolation at that point. This is illustrated in Fig. 3.1 where the reconstruction of a signal from its samples using the cardinal series is shown. The value of $x(t)$ at a point other than a sample location [*e.g.* $t = (m+\frac{1}{2})/2B$] is determined by all of the sample values.

3.2 Proofs

We now present three proofs of the conventional sampling theorem. The first is the one most commonly presented in texts. The second, due to Shannon, exposes the sampling theorem as the Fourier transform dual of the Fourier series. Finally, an eloquently compact form of proof due to Papoulis (1977) is presented.

3.2.1 Using Comb Functions

In this section we present the standard textbook proof of the sampling theorem. Since it can be nicely illustrated, the proof is quite instructive. It requires only an introductory knowledge of Fourier analysis.

In presenting this proof, we will repeatedly refer to Fig. 3.2 where five functions and their Fourier transforms are shown. In (a) is pictured a signal which, as is seen from its transform, is bandlimited with bandwidth B. The sampling is performed by multiplying the signal by the sequence of Dirac deltas shown in (b). The result, shown in (c) is

$$s(t) = x(t)\, 2B \operatorname{comb}(2Bt) \tag{3.5}$$

$$= \sum_{n=-\infty}^{\infty} x(\frac{n}{2B})\, \delta(t - \frac{n}{2B}). \tag{3.6}$$

Our goal is to recover $x(t)$ from $s(t)$ which is specified only by the signal's samples.

Let's examine what happens in the frequency domain when we sample. Multiplication in the time domain corresponds to convolution in the frequency domain. Since

$$2B \operatorname{comb}(2Bt) \longleftrightarrow \operatorname{comb}\left(\frac{u}{2B}\right),$$

we conclude that the transform of (3.5) is

$$S(u) = X(u) * \operatorname{comb}\left(\frac{u}{2B}\right).$$

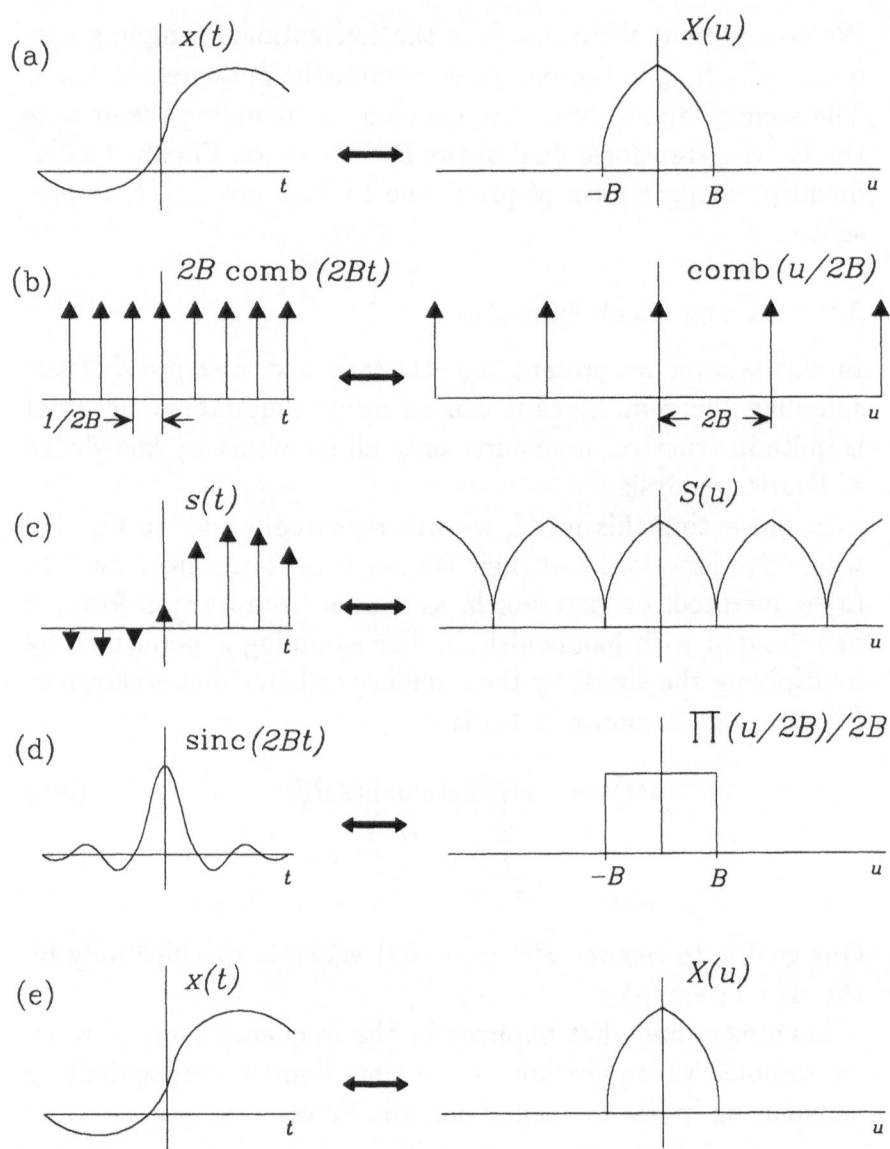

Figure 3.2: Illustration of Shannon's proof of the sampling theorem. Sampling is performed by multiplying lines (a)
and (b) in time to obtain line (c). The signal is regained by multiplying lines (c) and (d) in frequency
to obtain line (e).

Because convolving a function with $\delta(u-a)$ centers that function at a,

$$F(u) * \delta(u - a) = F(u - a),$$

we conclude that

$$S(u) = 2B \sum_{n=-\infty}^{\infty} X(u - 2nB). \tag{3.7}$$

These replications do not overlap. If $X(u)$ was not truly bandlimited or if the sampling rate were below $2B$, then the spectra would overlap. This phenomena is referred to as *aliasing*. Here, the low frequency components of the signal can still be regained although the high frequency components are irretrievably lost. Clearly, sampling can be performed at a rate greater than $2B$ without aliasing. The minimum sampling rate resulting in no aliasing (in this case $2B$) is referred to as the *Nyquist rate*. Sampling above the Nyquist rate can relax interpolation formula requirements (Section 4.1.1) and improve noise sensitivity (Section 5.1.1.2). If there are Dirac deltas in the signal's spectrum, sampling a bit above twice the signal's bandwidth may be required (Section 3.4).

Sampling is performed in Fig 3.2 by multiplying lines (a) and (b) in time to obtain line (c). Clearly, if there is no aliasing, multiplication of $S(u)$ on line (c) by the low pass filter on line (d) will result in the original spectrum, $X(u)$. That is

$$X(u) = S(u) \frac{1}{2B} \Pi\left(\frac{u}{2B}\right).$$

The corresponding time domain operation is convolution:

$$x(t) = s(t) * \mathrm{sinc}(2Bt).$$

Substituting (3.6) and evaluating gives the sampling theorem in (3.3).

3.2.2   Fourier Series Proof

Since $X(u)$ is identically zero for $|u| > B$, we can replicate it to form a periodic function in the frequency domain with period

$2B$. This periodic function can be expressed as a Fourier series. The result of the series for $|u| < B$ is $X(u)$. Using results in Sec. 2.2.1, we have

$$X(u) = \sum_{n=-\infty}^{\infty} c_n \, e^{-j\pi nu/B} \, \Pi\left(\frac{u}{2B}\right) \qquad (3.8)$$

where the Fourier coefficients are

$$\begin{aligned} c_n &= \frac{1}{2B} \int_{-B}^{B} X(u) \, e^{j\pi nu/B} \, du \\ &= \frac{1}{2B} x\left(\frac{n}{2B}\right). \end{aligned} \qquad (3.9)$$

Substituting into (3.8) and inverse transforming gives the sampling theorem series in (3.3) which, as we see here, is the Fourier transform dual of the Fourier series.

3.2.3   Papoulis' Proof

A quite eloquent proof of the sampling theorem begins with the Fourier series expansion of a periodic function with period $2B$ that is equal to $\exp(j2\pi ut)$ for $|u| < B$:

$$e^{j2\pi ut} = \sum_{n=-\infty}^{\infty} \text{sinc}(2Bt - n) \, e^{j2\pi(n/2B)u}. \qquad (3.10)$$

Substituting into the inversion formula in (3.2) gives

$$x(t) = \sum_{n=-\infty}^{\infty} \int_{-B}^{B} X(u) \, e^{j\pi nu/B} \, du \; \text{sinc}(2Bt - n).$$

Evaluating the integral by again using (3.2) gives the cardinal series in (3.3).

3.3   Properties

We now present some properties of the cardinal series.

3.3.1   Convergence

Although there are exceptions, the cardinal series generally converges uniformly. That is

$$\lim_{N \to \infty} |x(t) - x_N(t)| = 0 \qquad (3.11)$$

where the truncated cardinal series is

$$x_N(t) = \sum_{n=-N}^{N} x\left(\frac{n}{2B}\right) \text{sinc}(2Bt - n). \qquad (3.12)$$

The validity of (3.11) is obvious at the sample locations ($t = \frac{m}{2B}$) since contributions from adjacent points are zero. Uniform convergence is stronger than, say, the mean square convergence characteristic of Fourier series expansions of functions with discontinuities.

3.3.1.1   For Finite Energy Signals

We first will prove (3.11) for the case where $x(t)$ [and thus $X(u)$] has finite energy [Gallagher and Wise]. Note that

$$x(t) - x_N(t) = \int_{-B}^{B} [X(u) - X_N(u)] e^{j2\pi ut} \, du \qquad (3.13)$$

where

$$x_N(t) \longleftrightarrow X_N(u)$$
$$= \frac{1}{2B} \sum_{n=-N}^{N} x\left(\frac{n}{2B}\right) e^{-j\pi nu/B} \Pi\left(\frac{u}{2B}\right).$$

Schwarz's inequality can be written

$$\left| \int_{-\infty}^{\infty} A(u) C(u) \, du \right|^2 \leq \int_{-\infty}^{\infty} |A(u)|^2 \, du \int_{-\infty}^{\infty} |C(u)|^2 \, du. \qquad (3.14)$$

Application to (3.13) yields

$$|x(t) - x_N(t)|^2 \leq 2B \int_{-B}^{B} |X(u) - X_N(u)|^2 \, du. \qquad (3.15)$$

The right side approaches zero if $X_N \to X$ in the mean square sense. Since the limit of X_N is the Fourier series of X (see Sec.3.2.2) and the Fourier series displays mean square convergence (see Sec. 2.2.1), the right side of (3.15) tends to zero and our proof is complete.

3.3.1.2 For Bandlimited Functions with Finite Area Spectra

We now present an alternate proof of the cardinal series' uniform convergence for the case where $X(u)$ has finite area:

$$\int_{-B}^{B} |X(u)|\, du < \infty. \tag{3.16}$$

From Exercise 2.19, this constraint requires that $x(t)$ be bounded. Our proof will, for example, allow cardinal series representation for $J_0(2\pi t)$ and $x(t) = constant$ both of which, due to infinite energy, are excluded in the conditions of the previous proof.

We begin our proof by defining

$$e_N(u;t) = \sum_{n=-N}^{N} \text{sinc}(2Bt - n)\, e^{j\pi nu/B}. \tag{3.17}$$

From (3.10), we recognize that $e_N(u;t)$ is a truncated Fourier series of $\exp(j2\pi ut)$ for $|u| < B$. Since, for a fixed t, $\exp(j2\pi ut)$ is continuous for $|u| < B$, the Fourier series in (3.10) converges pointwise on this interval. Define the error magnitude

$$\varepsilon_N(u;t) = \left| e^{j2\pi ut} - e_N(u;t) \right|.$$

Let Δ_N denote the maximum (or supremum) of $\varepsilon_N(u;t)$ for $|u| < B$. Then we are guaranteed that

$$\lim_{N \to \infty} \Delta_N = 0. \tag{3.18}$$

Next, note that the truncated cardinal series in (3.12) can be written as

$$
\begin{aligned}
x_N(t) &= \sum_{n=-N}^{N} \int_{-B}^{B} X(u)\, e^{j\pi nu/B}\, du\, \text{sinc}(2Bt - n) \\
&= \int_{-B}^{B} X(u)\, e_N(u;t)\, du.
\end{aligned}
$$

Using the inequality

$$\left| \int Y(u)\,du \right| \le \int |Y(u)|\,du \tag{3.19}$$

we find that

$$|x(t) - x_N(t)| = \left| \int_{-B}^{B} X(u) \left[e^{j2\pi ut} - e_N(u;t) \right] du \right|$$

$$\le \int_{-B}^{B} |X(u)| e_N(u;t)\,du.$$

Since the integrand is positive,

$$|x(t) - x_N(t)| \le \Delta_N \int_{-B}^{B} |X(u)|\,du.$$

From (3.16), the integral is finite. Use of (3.18) results in (3.11) and the proof is complete.

The uniform convergence result for the cardinal series should not be surprising. Because of limited frequency constraints, bandlimited functions are inherently smooth. The sinc interpolation function is similarly smooth. There is thus no mechanism by which deviations such as Fourier series' *Gibb's phenomena* can occur (see Exercise 2.12).

3.3.2   Trapezoidal Integration

3.3.2.1   Of Bandlimited Functions

Clearly, if $x(t) \leftrightarrow X(u)$, then

$$X(0) = \int_{-\infty}^{\infty} x(t)\,dt.$$

Using (3.8) with (3.9), we thus conclude the integral of a band-limited function can be written directly in terms of its samples:

$$\int_{-\infty}^{\infty} x(t)\,dt = \frac{1}{2B} \sum_{n=-\infty}^{\infty} x\left(\frac{n}{2B} \right).$$

Thus, as illustrated in Fig. 3.3, trapezoidal integration of band-limited signals contains no error due to the piecewise linear approximation of the signal [Fig. 3.3b] if sampling is at or above the Nyquist rate. Since integration is over all of t, trapezoidal integration gives the same result as piecewise constant (rectangular) integration [Fig. 3.3c]

Accurate integration results from a signal's samples can also be obtained when sampling below the Nyquist rate. The details are left as an exercise at the end of the chapter.

3.3.2.2 Of Linear Integral Transforms

Consider numerical evaluation of the linear integral transform:

$$g(t) = \int_{-\infty}^{\infty} u(\tau) \, h(t; \tau) \, d\tau \qquad (3.20)$$

where $u(\tau)$ is the input, $g(t)$ is the transform and $h(t; \tau)$ is the transform kernel. Special cases are numerous and include correlation, convolution, and Laplace, Abel, Mellin, Hilbert and Hankel transforms [Bracewell].

One popular approach is to evaluate (3.20) by trapezoidal integration:

$$g(t) \approx g_\Delta(t) = \Delta \sum_{n=-\infty}^{\infty} u(n\Delta) \, h(t; n\Delta) \qquad (3.21)$$

where Δ is the input sampling interval. If the output is sampled, (3.21) can be expressed simply as a matrix–vector product.

We will show that by a simple alteration of the transform kernel, the expression in (3.21) can be made exact in the spirit of the sampling theorem [Marks (1981)]. Certain linear operations that can not be directly evaluated by use of (3.21) because of singularities can be evaluated through this sampling theorem characterization.

3.3.2.2.1.   Derivation of the Low Passed Kernel

Let $u(\tau)$ be bandlimited in the low pass sense with bandwidth B. Let $W > B$. Then $u(\tau)$ is unaffected by low-pass filtering:

$$u(\tau) = 2W \int_{-\infty}^{\infty} u(\eta) \, \text{sinc}[2W(\tau - \eta)] \, d\eta.$$

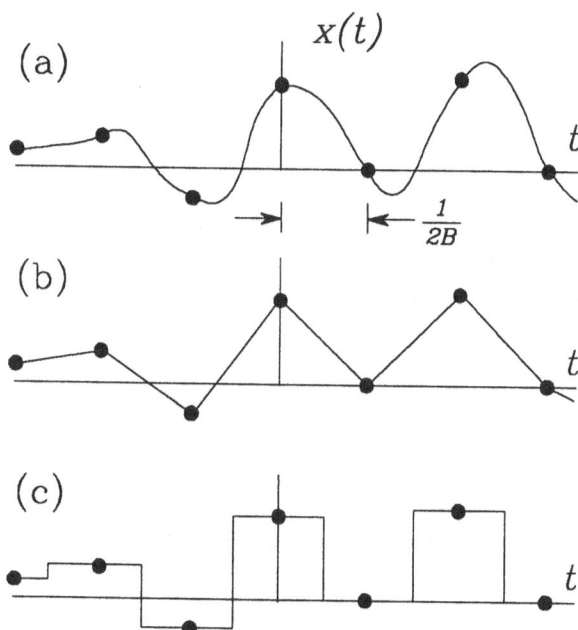

Figure 3.3: When sampling is performed at or above the Nyquist rate, the integration of $x(t)$ in (a) over all t gives the same result as integrating over (b) a piecewise linear connection of the samples (trapezoidal integration) and (c) a piecewise constant representation of the signal.

Substituting into (3.20) gives

$$g(t) = \int_{-\infty}^{\infty} u(\eta)\, k(t;\eta)\, d\eta \qquad (3.22)$$

where the *low-passed kernel* (LPK) is

$$k(t;\eta) = 2W \int_{-\infty}^{\infty} h(t;\tau)\,\text{sinc}[2W(\tau - \eta)]\, d\tau. \qquad (3.23)$$

Even though the kernel in (3.22) is altered, it yields the same result as in Eq. (3.20).

Since both the input and the LPK are bandlimited in η, they can be expressed by the cardinal series:

$$u(\eta) = \sum_{n=-\infty}^{\infty} u(n\Delta) \operatorname{sinc}\left(n - \frac{\eta}{\Delta}\right) \tag{3.24}$$

and

$$k(t;\eta) = \sum_{m=-\infty}^{\infty} k(t; m\Delta) \operatorname{sinc}\left(m - \frac{\eta}{\Delta}\right) \tag{3.25}$$

where the input sampling interval must be chosen such that

$$\Delta \leq \frac{1}{2W} \leq \frac{1}{2B}.$$

Substituting Eqs.(3.24)and (3.25) into Eq.(3.22) gives

$$
\begin{aligned}
g(t) &= \sum_{n=-\infty}^{\infty} \sum_{m=-\infty}^{\infty} u(n\Delta) k(t; m\Delta) \int_{-\infty}^{\infty} \operatorname{sinc}\left(n - \frac{\eta}{\Delta}\right) \operatorname{sinc}(m - \frac{\eta}{\Delta}) d\eta \\
&= \Delta \sum_{n=-\infty}^{\infty} u(n\Delta) k(t; n\Delta). \tag{3.26}
\end{aligned}
$$

This is the desired result. Comparing it with Eq.(3.21), we conclude that the inaccuracy that is due to trapezoidal integration can be totally eliminated if the LPK is used in lieu of the original kernel.

3.3.2.2.2.   Example Transforms

To illustrate use of (3.26), we now present example applications for the cases of Laplace and Hilbert transformation.

Laplace Transform

The (unilateral) Laplace transform can be written as

$$g(t) = \int_0^\infty u(\tau) e^{-t\tau} d\tau.$$

Comparing with (3.20) and (3.22), we have

$$h(t; \tau) = e^{-t\tau} \mu(\tau)$$

and

$$k(t; \eta) = \int_0^\infty e^{-t\tau} \operatorname{sinc}(\tau - \eta) \, d\tau$$

where $\mu(\cdot)$ denotes the unit step and we have chosen $2W = 1$.

EXAMPLE:

Consider the Laplace transform of $u(t) = \operatorname{sinc}(t)$. We evaluate the resulting Laplace transform integral in three ways.

- DIRECTLY.
  The Laplace integral becomes

$$g(t) = \int_0^\infty \operatorname{sinc}(\tau) e^{-t\tau} \, d\tau. \tag{3.27}$$

To evaluate this integral, consider

$$\gamma = \int_0^\infty \frac{\sin(a\tau)}{\tau} e^{-t\tau} \, d\tau.$$

Clearly

$$\frac{\partial \gamma}{\partial a} = \int_0^\infty \cos(a\tau) e^{-t\tau} \, d\tau$$

$$= \frac{t}{t^2 + a^2}.$$

Thus, using the boundary condition that $a = 0 \longrightarrow \gamma = 0$, we conclude that $\gamma = \arctan(\frac{a}{t})$, and the true Laplace transform of $\operatorname{sinc}(\tau)$ is

$$g(t) = \frac{1}{\pi} \arctan\left(\frac{\pi}{t}\right).$$

- USING TRAPEZOIDAL INTEGRATION
  Applying trapezoidal integration to the integral in (3.27)

with a step size of Δ gives [1]

$$g(t) \approx g_\Delta(t) \;=\; \frac{\Delta}{2} + \Delta \sum_{n=1}^{\infty} \text{sinc}(n\Delta)\, e^{-tn\Delta}$$

$$= \frac{\Delta}{2} + \frac{1}{\pi}\Im\sigma \qquad (3.28)$$

where $\Im$ is the imaginary operator,

$$\sigma = \sum_{n=1}^{\infty} \frac{1}{n} Z^n,$$

and $Z = \exp[-(t - j\pi)\Delta]$. Since

$$\frac{\partial\sigma}{\partial Z} \;=\; \sum_{m=0}^{\infty} Z^m$$

$$= \frac{1}{1 - Z}$$

we conclude, since $Z = 0 \longrightarrow \sigma = 0$, that

$$\sigma = -\ln(1 - Z).$$

For $Z = |Z|\, e^{j\,\text{arg}(Z)}$, recall that $\Im \ln Z = \text{arg}(Z)$. Thus, after some substitution, (3.28) becomes

$$g_\Delta(t) = \frac{\Delta}{2} + \frac{1}{\pi} \arctan\left[\frac{\sin(\pi\Delta)}{e^{t\Delta} - \cos(\pi\Delta)} \right]. \qquad (3.29)$$

Note that, as $\Delta \longrightarrow 0$,

$$\frac{\sin(\pi\Delta)}{e^{t\Delta} - \cos(\pi\Delta)} \longrightarrow \frac{\pi}{t}.$$

Thus, as we would expect,

$$\lim_{\Delta \to 0} g_\Delta(t) = g(t).$$

---

[1]The $\frac{\Delta}{2}$ term is replaced by Δ for rectangular integration.

A plot of $g_\Delta(t)$ for various $\Delta's$ is shown in Figure 3.4 along with $g(t)$. Recall that for $\Delta = 1$, the LPK approach, in the spirit of the cardinal series, gives exact results.

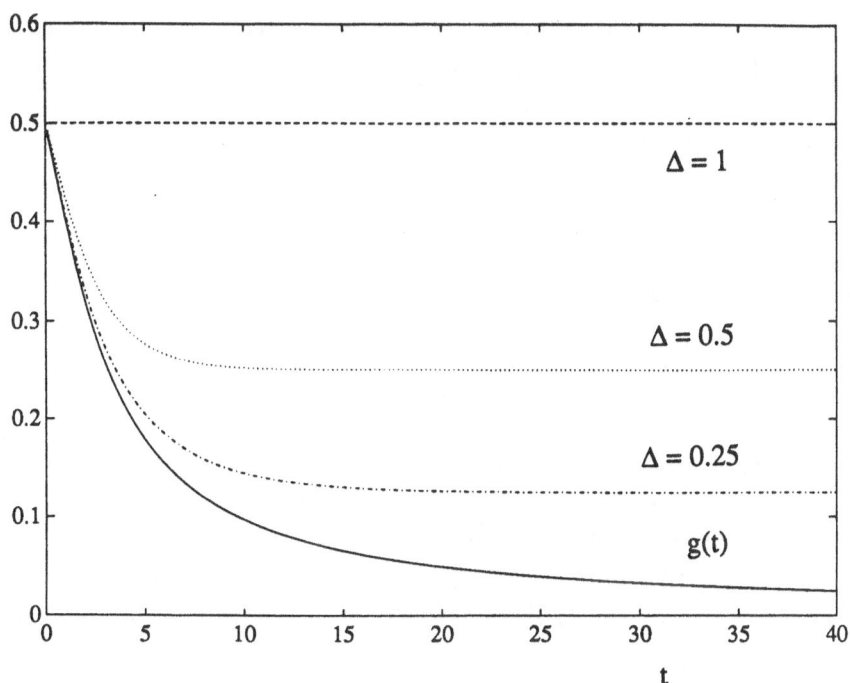

Figure 3.4: Evaluation of the Laplace transform of sinc(t) by trapezoidal integration. Shown is $g_\Delta(t)$ for $\Delta = 1, 0.5$ and 0.25. Use of t he low passed kernel will converge to the desired $g(t)$ for step sizes, Δ, less than or equal to one.

- USING THE LPK

  For $u(\tau) = \text{sinc}(\tau)$ and $2W = 2B = 1/\Delta = 1$, the LPK expression in (3.26) becomes

$$g(t) = k(t; 0)$$

where we have used the property that $\text{sinc}(n) = \delta[n]$. Using (3.23) with $\eta = 0$ gives

$$g(t) = \int_{-\infty}^{\infty} h(t; \tau) \, \text{sinc}(\tau) \, d\tau$$

which is the integral we wished to evaluate in the first place. The LPK, when applied to $\text{sinc}(t)$, therefore reduces to an identity for Laplace transformation or, for that matter, any other linear transform.

Hilbert Transform

The Hilbert transform

$$g(t) = -\frac{1}{\pi} \int_{-\infty}^{\infty} \frac{u(\tau) \, d\tau}{t - \tau} \tag{3.30}$$

cannot be accurately evaluated by direct trapezoidal integration because of the singularity at $\tau = t$. Fourier–transform analysis, rather, is commonly used. We will now show, however, that through application of the LPK, an accurate matrix-vector characterization of the Hilbert transform is possible. From (3.30)

$$h(t; \tau) = \frac{-1}{\pi(t - \tau)}.$$

If we choose $W = B$, the corresponding LPK is

$$
\begin{aligned}
k(t; \eta) &= \int_{-B}^{B} \left[-\frac{1}{\pi} \int_{-\infty}^{\infty} \frac{\exp(-j2\pi u \tau)}{t - \tau} \, d\tau \right] e^{j2\pi u \eta} \, du \\
&= -j \int_{-B}^{B} \text{sgn}(u) \, e^{-j2\pi u(t-\eta)} \, du \\
&= -2B \sin \pi B(t - \eta) \, \text{sinc} B(t - \eta). \tag{3.31}
\end{aligned}
$$

A further simplification arises after we note that Hilbert transformation is a shift–invariant operation. Thus, if $u(\tau)$ is band-limited, so then is the output, $g(t)$. This being true, we need to know g only at the points where $t = m/2B$. Substituting (3.31) into (3.26) with $\Delta = 1/2B$ gives

$$g(m\Delta) = \sum_{n=-\infty}^{\infty} u(n\Delta) \sin\left[\frac{\pi}{2}(m-n)\right] \text{sinc}\left[\frac{1}{2}(m-n)\right].$$

Noting that every other term is zero yields the final desired result.

$$g(m\Delta) = \frac{-2}{\pi} \sum_{m-n \text{ odd}} \frac{u(n\Delta)}{m-n}.$$

This discrete convolution version of the Hilbert transform contains no singularities and is exact for all band-limited inputs.

3.3.2.3 Parseval's Theorem for the Cardinal Series

The energy of a signal is

$$E = \int_{-\infty}^{\infty} |x(t)|^2 \, dt.$$

For a bandlimited signal, we substitute the cardinal series in (3.3) and write :

$$E = \sum_{n=-\infty}^{\infty} \sum_{m=-\infty}^{\infty} x\left(\frac{n}{2B}\right) x^*\left(\frac{m}{2B}\right) I_{nm}$$

where

$$I_{nm} = \int_{-\infty}^{\infty} \text{sinc}(2Bt - n)\,\text{sinc}(2Bt - m) \, dt.$$

Using the power theorem

$$
\begin{aligned}
I_{nm} &= \frac{1}{(2B)^2} \int_{-B}^{B} e^{-j\pi(n-m)u/B} \, du \\
&= \frac{1}{2B} \delta[n-m].
\end{aligned}
$$

This gives Parseval's theorem for the sampling theorem:

$$E = \frac{1}{2B} \sum_{n=-\infty}^{\infty} \left| x\left(\frac{n}{2B}\right) \right|^2. \tag{3.32}$$

The signal's energy can thus be determined by summing the square of the magnitude of each sample.

3.3.3   The Time–Bandwidth Product

The cardinal series requires an infinite number of samples. Since all bandlimited functions are analytic, they cannot be identically zero over any finite subinterval (except for the degenerate case $x(t) \equiv$ zero). Thus, the number of nonzero samples taken from almost every band-limited function is finite. The only exceptions are signals that can be expressed as the sum of a finite number of uniformly spaced sinc functions.

We can, however, have a "good" representation of the function using a finite number of samples. If a signal has either finite area or finite energy, it must asymptotically approach zero at $t = \pm\infty$. In such cases, there is always an interval of duration T outside of which the samples are negligibly small. If we sample over this interval at the Nyquist rate, $2B$, then a total of

$$S = 2BT$$

samples are needed to characterize the signal. This quantity, the *time–bandwidth product*, measures the number of degrees of freedom of the signal. It has also been termed the *Shannon number* [DiFrancia].

Choice of T is dictated by the *truncation error* one can tolerate. This topic is treated in Section 5.3.

3.4   Application to Spectra Containing Distributions

The cardinal series is applicable in certain cases where $X(u)$ contains distributions such as the Dirac delta. Indeed $x(t) = 1$ is bandlimited in the sense of (3.1) for all $B > 0$ and falls into the category of bandlimited signals with finite area spectra. The corresponding cardinal series is

$$\sum_{n=-\infty}^{\infty} \text{sinc}(2Bt - n) = 1. \tag{3.33}$$

Similarly, $\cos(2\pi Bt - \phi)$ has a transform containing two Dirac delta functions at $u = \pm B$. To insure both deltas are contained in the replicated sample spectrum, the sampling rate, $\frac{1}{T}$, must exceed $2B$ and

$$\cos(2\pi Bt - \phi) = \sum_{n=-\infty}^{\infty} \cos(\pi rn - \phi) \operatorname{sinc}\left(\frac{t}{T} - n\right)$$

where

$$r = 2BT < 1. \tag{3.34}$$

If $r = 1$, the bandwidth interval begins and ends at delta function locations. We are confronted with the unanswerable question of what percentage of each delta should be included in the bandwidth interval. Requiring (3.34) to be a strict inequality avoids this problem. Note, also, for $r = 1$, it is possible to have every sample be zero. The resulting interpolation clearly would be identically zero and therefore incorrect.

A distribution whose inverse transform does not have a valid cardinal series is the *unit doublet*–the derivative of the Dirac delta:

$$\delta^{(1)}(u) = \left(\frac{d}{du}\right)\delta(u).$$

From the dual of the derivative theorem,

$$-j2\pi t \longleftrightarrow \delta^{(1)}(u).$$

Thus, $x(t) = t$ is bandlimited in the sense of (3.1). Using the form of the cardinal series in (3.4), we ask the question:

$$t \overset{?}{=} \frac{1}{\pi} \sin(2\pi Bt) \sum_{n=-\infty}^{\infty} (-1)^n \frac{(n/2B)}{(2Bt - n)}.$$

For a fixed $t \neq m/2B$, the answer is clearly "no" since the nth term in the sum approaches $(-1)^{n+1}$ – an oscillatory and thus divergent series. The truncated cardinal series thus does not asymptotically approach the desired value.

An alteration of the cardinal series to make expansions of such singularities valid has been proposed by Zemanian.

3.5 Application to Bandlimited Stochastic Processes

A real wide sense stationary stochastic process, $f(t)$, is said to be band-limited if its spectral density obeys

$$S_f(u) = S_f(u)\, \Pi\left(\frac{u}{2B}\right).$$

As a consequence, the autocorrelation is a bandlimited function. We will use this observation to show mean square convergence of the cardinal series for $f(t)$. Specifically, define

$$\hat{f}(t) = \sum_{n=-\infty}^{\infty} f\left(\frac{n}{2B}\right) \operatorname{sinc}(2Bt - n).$$

Then $\hat{f}(t)$ is equal to $f(t)$ in the mean square sense:

$$E\left[\{f(t) - \hat{f}(t)\}^2\right] = 0. \tag{3.35}$$

Our proof begins by expansion of the mean square error expression:

$$
\begin{aligned}
E[\{f(t) - \hat{f}(t)\}^2] = \overline{f^2} & \\
+ \sum_{n=-\infty}^{\infty} \sum_{m=-\infty}^{\infty} R_f & \left(\frac{n - m}{2B}\right) \operatorname{sinc}(2Bt - n)\operatorname{sinc}(2Bt - m) \\
- 2 \sum_{n=-\infty}^{\infty} R_f & \left(t - \frac{n}{2B}\right) \operatorname{sinc}(2Bt - n) \\
= \overline{f^2} + T_2 + T_3. &
\end{aligned}
\tag{3.36}
$$

In the second term, we make the variable substitution $k = n - m$:

$$T_2 = \sum_{k=-\infty}^{\infty} R_f\left(\frac{k}{2B}\right) \sum_{n=-\infty}^{\infty} \operatorname{sinc}(2Bt - n + k)\operatorname{sinc}(2Bt - n).$$

Application of the cardinal series yields, for arbitrary τ,

$$\sum_{n=-\infty}^{\infty} \operatorname{sinc}(2B\tau - n)\,\operatorname{sinc}(2Bt - n) = \operatorname{sinc} 2B(t - \tau). \tag{3.37}$$

Using the result of substituting $\tau = t + \frac{k}{2B}$ in (3.37) gives

$$
\begin{aligned}
T_2 &= \sum_{k=-\infty}^{\infty} R_f\left(\frac{k}{2B}\right) \delta[k] \\
&= \overline{f^2}.
\end{aligned}
\tag{3.38}
$$

To evaluate the third term, we recognize that $R_f(\tau - t)$ is bandlimited and can be written as a cardinal series:

$$
R_f(\tau - t) = \sum_{n=-\infty}^{\infty} R_f\left(\tau - \frac{n}{2B}\right) \text{sinc}(2Bt - n).
$$

Evaluation at $t = \tau$ gives

$$
T_3 = -2\,\overline{f^2}.
$$

Substituting this and (3.38) into (3.36) yields (3.35) and our proof is complete.

3.6   Exercises

3.1. The derivation of the Poisson sum formula closely parallels Shannon's proof of the sampling theorem. Starting with the Fourier dual of the Poisson sum formula, derive the sampling theorem series.

3.2. The integral of a bandlimited signal

$$I = \int_{-\infty}^{\infty} x(t)\, dt$$

can be determined from signal samples taken *below* the Nyquist rate. Find I from $\{x(nT)|n = 0, \pm 1, \pm 2, \ldots\}$ when $B < 1/T < 2B$ and $X(u) = X(u)\, \Pi(\frac{u}{2B})$.

3.3. Show that, for any real α, if $x(t)$ can be expressed by the cardinal series, then

$$x(t) = \sum_{n=-\infty}^{\infty} x\left(\frac{n}{2B} + \alpha\right) \operatorname{sinc}[2B(t - \alpha) - n].$$

3.4. Investigate application of the cardinal series to a signal whose spectrum is an mth–*let*:

$$\delta^{(m)}(u) = \left(\frac{d}{du}\right)^m \delta(u).$$

When, if ever, is the cardinal series applicable here? Assume $m > 1$.

3.5. Let $y(t)$ be any well behaved (not necessarily bandlimited) function. We sample $y(t)$ at a rate of $\frac{1}{T}$ and, in the spirit of (3.5), define

$$\hat{s}(t) = y(t) \frac{1}{T} \operatorname{comb}\left(\frac{t}{T}\right).$$

Show that

$$T \int_{-1/T}^{1/T} \hat{S}(u)\, du = \int_{-\infty}^{\infty} Y(u)\, du.$$

3.6. Let $x(t)$ and $y(t)$ denote two finite energy bandlimited functions with bandwidth B. How is the series

$$\sum_{n=-\infty}^{\infty} x\left(\frac{n}{2B}\right) y^*\left(\frac{n}{2B}\right)$$

related to

$$\int_{-\infty}^{\infty} x(t)\, y^*(t)\, dt?$$

3.7. The spectrum of a real signal is hermetian (*i.e.* it is equal to the conjugate of its transpose). Thus, if we know the spectrum for positive frequencies, we know it for negative frequencies. Visualize setting the negative frequency components to zero and shifting the remaining portion of the spectrum to be centered about the origin. Clearly, we have reduced the bandwidth by a factor of one half yet have lost no information. Explain, however, why the sampling density of this new signal is the same as that required by the original.

3.8. Apply the low passed kernel technique to Fourier inversion of a bandlimited function. Let $f(t)$ have a bandwidth of B and use the Fourier kernel

$$h(u; t) = \exp(-j2\pi u t).$$

Comment on the usefulness of the result.

REFERENCES

R.N. Bracewell, **The Fourier Transform and Its Applications, 2nd edition, Revised**, McGraw-Hill, New York, 1986.

G. T. diFrancia "Degrees of freedom of an image", *Journal of the Optical Society of America*, vol. 59, pp.799-804 (1969).

N.C. Gallagher Jr. and G.L. Wise "A representation for bandlimited functions", *Proc. IEEE*, vol. 63, p.1624 (1975).

R.J. Marks II "Sampling theory for linear integral transforms", *Optics Letters*, vol. 6, pp.7-9 (1981).

A. Papoulis, **Signal Analysis**, McGraw-Hill, New York, 1977.

A.H. Zemanian, **Distribution Theory and Transform Analysis**, McGraw-Hill, New York, 1965.

A.H. Zemanian, **Generalized Integral Transforms, Interscience**, New York, 1968.

4

Generalizations of the Sampling Theorem

There have been a number of significant generalizations of the sampling theorem. Some are straightforward variations on the fundamental cardinal series. Oversampling, for example, results in dependent samples and allows much greater flexibility in the choice of interpolation functions. In Chapter 5, we will see that it can also result in better performance in the presence of sample data noise.

Bandlimited signal restoration from samples of various filtered versions of the signal is the topic addressed in Papoulis' generalization of the sampling theorem. Included as special cases are recurrent nonuniform sampling and simultaneously sampling a signal and one or more of its derivatives.

Kramer (1959) generalized the sampling theorem to signals that were bandlimited in other than the Fourier sense. Specifically, integral kernels other than $\exp(j2\pi ut)$ are allowed.

We also demonstrate that the cardinal series is a special case of Lagrangian polynomial interpolation.

4.1 Generalized Interpolation Functions

There are a number of functions other than the sinc which can be used to weigh a signal's samples in such a manner as to uniquely characterize the signal. Use of these *generalized interpolation functions* allows greater flexibility in dealing with sampling theorem type characterizations.

4.1.1   Oversampling

If a bandlimited signal has bandwidth B, then it can also be considered to have bandwidth $W \geq B$. Thus,

$$x(t) = \sum_{n=-\infty}^{\infty} x\left(\frac{n}{2W}\right) \text{sinc}(2Wt - n). \qquad (4.1)$$

Note, however, since

$$x(t) = x(t) * 2B \text{ sinc}(2Bt)$$

we can write:

$$
\begin{aligned}
x(t) &= \sum_{n=-\infty}^{\infty} x\left(\frac{n}{2W}\right) \text{sinc}(2Wt - n) * 2B \text{ sinc}(2Bt) \\
&= r \sum_{n=-\infty}^{\infty} x\left(\frac{n}{2W}\right) \text{sinc}(2Bt - rn) \qquad (4.2)
\end{aligned}
$$

where the *sampling rate parameter* is

$$r = B/W \leq 1.$$

Equation (4.2) reduces to the conventional cardinal series for $r = 1$. In the next chapter, we will see that oversampling can be used to reduce interpolation noise level.

4.1.1.1   Sample Dependency

When a bandlimited signal is oversampled, its samples become dependent. Indeed, in this section we will show that in the absence of noise, any finite number of lost samples can be regained from those remaining. First, using clear geometrical arguments, we will illustrate the feasibility of lost sample recovery. Then, an alternate expression with better convergence properties will be derived.

4.1.1.1.1.   Restoring a Single Lost Sample

Consider a bandlimited signal $x(t)$ and its spectrum as shown at the top of Fig.4.1. From (3.5) and (3.7) the spectrum of the signal of samples

$$s(t) = x(t) \; 2W \text{ comb}(2Wt) \qquad (4.3)$$

is

$$S(u) = 2W \sum_{n=-\infty}^{\infty} X(u - 2nW). \qquad (4.4)$$

As is shown in Fig. 4.1, there are intervals identically equal to zero in $S(u)$ when we oversample.

In Fig. 4.1a, the sample of $x(t)$ at the origin has been set to zero. We can view this as the subtraction of $x(0)\delta(t)$ in Fig. 4.1c from $s(t)$

$$\hat{s}(t) = s(t) - x(0)\,\delta(t).$$

Given $\hat{s}(t)$, we can regain $x(0)$. Indeed, transforming gives

$$\hat{S}(u) = S(u) - x(0). \qquad (4.5)$$

Since $S(u) \equiv 0$ on the interval $B < |u| < 2W - B$,

$$\hat{S}(u) = -x(0) \quad ; \quad B < |u| < 2W - B.$$

This is illustrated in Fig. 4.1d. An appropriate point to sample in this interval is $u = W$. Thus:

$$x(0) = -\hat{S}(W). \qquad (4.6)$$

Therefore, the lost sample at the origin can be regained from those remaining.

Samples from oversampled signals display an interesting zero sum property. Writing(4.3) in terms of delta functions and transforming gives

$$S(u) = \sum_{n=-\infty}^{\infty} x\left(\frac{n}{2W}\right) e^{-j\pi nu/W}.$$

Since $S(W) = 0$, we conclude that

$$\sum_{n=-\infty}^{\infty} (-1)^n\, x\left(\frac{n}{2W}\right) = 0 \quad ; \quad r < 1. \qquad (4.7)$$

An alternate formula for restoring a lost sample (with better convergence properties) results directly from inspection of (4.2). Note that, unlike the conventional ($r = 1$) cardinal series, (4.2)

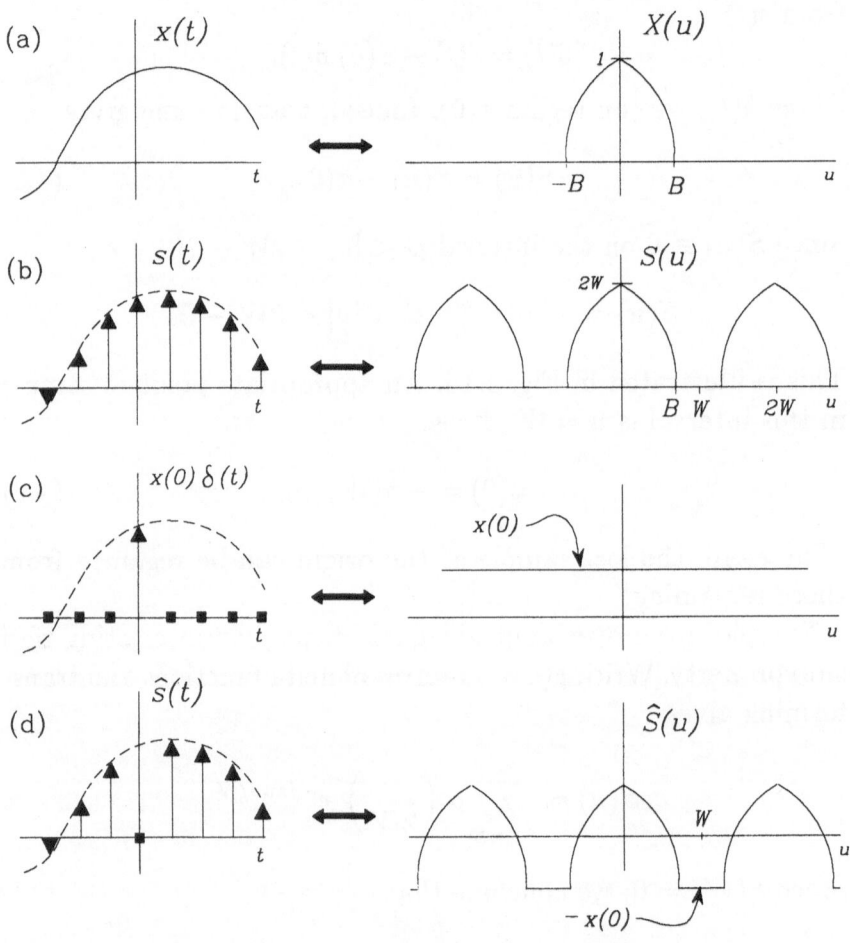

Figure 4.1: Geometrical illustration of restoring a single lost sample from an oversampled bandlimited signal.

does not reduce to an identity when $t = n/2W$. For example, at $t = 0$ we have

$$x(0) = r \sum_{n=-\infty}^{\infty} x\left(\frac{n}{2W}\right) \operatorname{sinc}(rn). \tag{4.8}$$

There is thus a dependency among samples. Indeed, isolating the $n = 0$ term in (4.8) and solving for $x(0)$ gives

$$x(0) = \frac{r}{1-r} \sum_{n \neq 0} x\left(\frac{n}{2W}\right) \operatorname{sinc}(rn). \tag{4.9}$$

The sample at the origin is thus completely specified by the remaining samples if $r < 1$. The convergence here is better than in (4.7) due to the $1/n$ decay of the summand from the sinc term. Equation(4.7), on the other hand, does not require knowledge of r.

From the cardinal series, we have

$$x(t) = x(0) \operatorname{sinc}(2Wt) + \sum_{n \neq 0} x\left(\frac{n}{2W}\right) \operatorname{sinc}(2Wt - n).$$

Substituting (4.9) and simplifying gives an interpolation formula not requiring knowledge of the sample at the origin:

$$x(t) = \sum_{n \neq 0} x(\frac{n}{2W})[\operatorname{sinc}(2Wt - n) + \frac{r}{1-r} \operatorname{sinc}(rn) \operatorname{sinc}(2Wt)].$$
$$\tag{4.10}$$

Both sides can be low pass filtered to give an alternate expression (see Exercise 4.20). The noise sensitivity and truncation error for this interpolation is explored in Chapter 5.

4.1.1.1.2.   Restoring M Lost Samples

This single sample restoration result can be generalized to restoring an arbitrarily large but finite number of lost samples. Let $\mathcal{M}$ denote a set of M integers corresponding to the locations of M lost samples. From the data set $\{x\left(\frac{n}{2W}\right) \mid n \notin \mathcal{M}\}$ we wish to find $\{x(\frac{n}{2W}) \mid n \in \mathcal{M}\}$.

To do this, we write (4.2) as

$$x(t) = r \left[\sum_{n \in \mathcal{M}} + \sum_{n \notin \mathcal{M}} \right] x \left(\frac{n}{2W} \right) \operatorname{sinc}(2Bt - rn). \qquad (4.11)$$

Evaluating this expression at the M points $\{t = \frac{m}{2W} \mid m \in \mathcal{M}\}$ and rearranging gives

$$\sum_{n \in \mathcal{M}} x \left(\frac{n}{2W} \right) \{\delta[n-m] - r \operatorname{sinc}[r(n-m)]\} = g(\frac{m}{2W}) \ ; \ m \in \mathcal{M}$$
$$(4.12)$$

where

$$g(t) = r \sum_{n \notin \mathcal{M}} x \left(\frac{n}{2W} \right) \operatorname{sinc}(2Bt - rn) \qquad (4.13)$$

can be computed from the known samples.

Equation (4.12) consists of M equations and M unknowns. In matrix form

$$\mathbf{H} \, \vec{x} = \vec{g}$$

where $\{g_m = g(\frac{m}{2W}) \mid m \in \mathcal{M}\}$, $\{x_n = x \left(\frac{n}{2W} \right) \mid n \in \mathcal{M}\}$ and $\mathbf{H} = \mathbf{I} - \mathbf{S}$ where $\mathbf{S}$ has elements

$$\{s_{nm} = s_{n-m} = r \operatorname{sinc}[r(n-m)] \mid (n,m) \in \mathcal{M} \times \mathcal{M}\}.$$

Clearly, we can determine the lost samples if $\mathbf{H}$ is not singular. It is when $r = 1$.

4.1.1.1.3. Direct Interpolation From M Lost Samples

Here we address direct generation of $x(t)$ from the samples $\{x(n/2W) \mid n \notin \mathcal{M}\}$. Using (4.13) the solution of (4.12) can be written as

$$
\begin{aligned}
x \left(\frac{q}{2W} \right) &= \sum_{p \in \mathcal{M}} a_{pq} \, g \left(\frac{p}{2W} \right) \\
&= r \sum_{n \notin \mathcal{M}} x \left(\frac{n}{2W} \right) \sum_{p \in \mathcal{M}} a_{pq} \operatorname{sinc}[r(p-n)]; \\
& \qquad\qquad q \in \mathcal{M}
\end{aligned}
\qquad (4.14)
$$

where $\{a_{pq} \mid (p,q) \in \mathcal{M} \times \mathcal{M}\}$ are elements of the inverse of $\mathbf{H}$. The cardinal series can be written:

$$x(t) = \sum_{n \notin \mathcal{M}} x\left(\frac{n}{2W}\right) \text{sinc}(2Wt-n) + \sum_{q \in \mathcal{M}} x\left(\frac{q}{2W}\right) \text{sinc}(2Wt-q).$$

Substituting (4.14) gives

$$x(t) = \sum_{n \notin \mathcal{M}} x\left(\frac{n}{2W}\right) k_n(2Wt) \tag{4.15}$$

where the interpolation function is

$$k_n(t) = \text{sinc}(t-n) + r \sum_{p \in \mathcal{M}} \sum_{q \in \mathcal{M}} a_{pq} \text{sinc}[r(n-p)] \text{sinc}(t-q). \tag{4.16}$$

Alternately, we can pass (4.15) through a filter unity for $|u| < B$ and zero elsewhere. The result is

$$x(t) = \sum_{n \notin \mathcal{M}} x\left(\frac{n}{2W}\right) k_n^{(r)}(2Bt) \tag{4.17}$$

and

$$k_n^{(r)}(2Bt) = k_n(2Wt) * 2B\,\text{sinc}(2Bt).$$

It is straightforward to show that

$$k_n^{(r)}(t) = r\text{sinc}(t-rn) + r^2 \sum_{p \in \mathcal{M}} \sum_{q \in \mathcal{M}} a_{pq}\text{sinc}(r(n-p))\text{sinc}(t-rq). \tag{4.18}$$

For $\mathcal{M}$ empty, (4.15) reduces to the cardinal series and (4.18) to the oversampled restoration formula in (4.8).

In the absence of noise we can, in general, restore an arbitrarily large number of lost samples if $r < 1$. In Chapter 5, we demonstrate that restoration becomes more and more unstable as M increases and r approaches one. The algorithm is extended to higher dimensions in Chapter 6.

4.1.1.2   Relaxed Interpolation Formulae

Here we show oversampled signals can tolerate rather significant perturbations in the interpolation function. Consider (4.4). If

$x(t)$ has a bandwidth of W, then the replicated spectra will be separated as is shown in Fig.4.2. Define

$$K_r(u) = \begin{cases} \frac{1}{2W} & ; \quad |u| < B \\ \text{anything convenient} & ; \quad B < |u| < 2W - B \end{cases}$$

$$(4.19)$$

where the subscript is for "relaxed." Clearly, since $S(u) = 0$ for $B < |u| < 2W - B$.

$$X(u) = S(u)\, K_r(u).$$

Inverse transforming gives

$$\begin{aligned} x(t) &= s(t) * k_r(t) \\ &= \sum_{n=-\infty}^{\infty} x\left(\frac{n}{2W}\right) k_r\left(t - \frac{n}{2W}\right) \end{aligned} \qquad (4.20)$$

where we have used (4.3). Both the cardinal series and (4.2) are subsumed in this expression. Also, in practice, we can relax the roll-off in the spectrum of the generalized interpolation function, $k_r(t)$, with no error cost in restoration (in the absence of data noise and truncation error).

4.1.2   Criteria for Generalized Interpolation Functions

Let $k(t)$ now be an arbitrary function and define

$$g(t) = \sum_{n=-\infty}^{\infty} f(\frac{n}{2B})\, k(2Bt - n). \qquad (4.21)$$

If $f(t)$ has a bandwidth of B, under what condition can we recover $f(t)$ from $g(t)$?
    Transforming (4.21) gives

$$\begin{aligned} G(u) &= \frac{1}{2B} \sum_{n=-\infty}^{\infty} f(\frac{n}{2B})\, e^{-j\pi n u/B}\, K(\frac{u}{2B}) \\ &= \sum_{n=-\infty}^{\infty} F(u - 2nB)\, K(\frac{u}{2B}) \end{aligned}$$

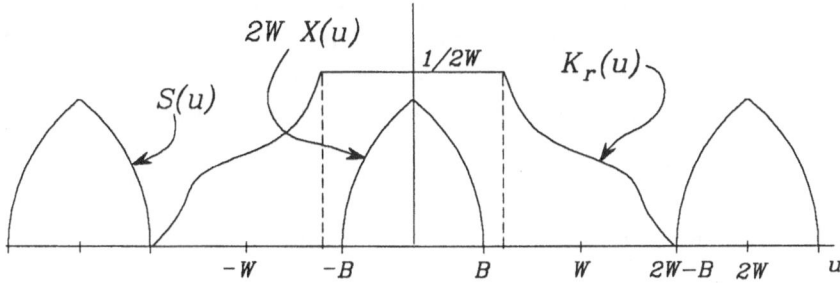

Figure 4.2: Replicated spectra for an oversampled signal allows flexibility in interpolation function choice.

where we have used the Fourier dual of the Poisson sum formula. The function, $G(u)$, is recognized as the replicated signal spectrum weighted by $K(u/2B)$. Define the transfer function

$$H(u) = \frac{\Pi(u)}{K(u)}. \tag{4.22}$$

Then the signal spectrum can be regained by

$$F(u) = G(u) \, H(\frac{u}{2B}).$$

Thus, $f(t)$ can be generated by passing $g(t)$ through a filter with impulse response $h(t)$. The cardinal series is the special case when $H(u) = \Pi(u)$.

Clearly, $H(u)$ does not exist if $K(u)$ is identically zero over any subinterval of $|u| < 1/2$. If $K(u)$ passes through zero for $|u| < B$, then restoration is still possible but is many times ill posed. (This is discussed in detail in Chapter 5). A sufficient condition for (well-posed) recovery is that $H(u)$ be bounded.

4.1.2.1   Interpolation Functions

The function $k(t)$ in (4.21) is said to be an *interpolation function* if the resulting interpolation passes through the samples. For (4.21), this is equivalent to requiring that

$$k(n) = \delta[n]. \tag{4.23}$$

This condition assures that the resulting interpolation passes through the sample points. Specifically, for $t = m/2B$, (4.21) reduces to an identity when (4.23) is valid.

Some commonly used interpolation functions follow:

(a) For $k(t) = \mathrm{sinc}(t)$, (4.21) becomes the cardinal series.

(b) Consider the interpolation function

$$k(t) = \Pi(t).$$

The resulting interpolation is referred to as a zeroth order *sample and hold*. An example is shown in Fig.4.3. To restore $f(t)$ from this interpolation, we pass the zero order sample and hold data through a filter with frequency response $H(u/2B)$ where

$$H(u) = \frac{\Pi(u)}{\mathrm{sinc}(u)}.$$

Note that $H(u)$ is bounded.

(c) *Piecewise linear interpolation* uses the interpolation function

$$k(t) = \Lambda(t).$$

The result, as is shown in Fig.4.3., is that the sample points are linearly connected. The signal $x(t)$ can be regained by passing this waveform through a filter with frequency response $H(u/2B)$ where

$$H(u) = \frac{\Pi(u)}{\mathrm{sinc}^2(u)}. \tag{4.24}$$

Again, $H(u)$ is bounded.

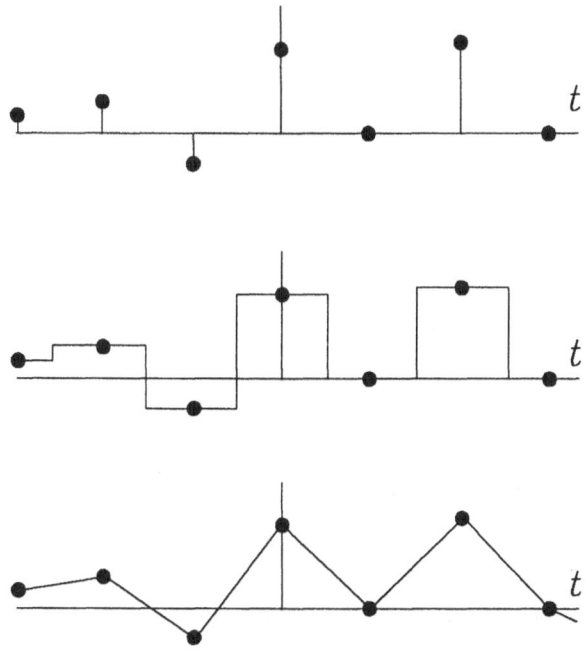

Figure 4.3: Sample data (top) with zeroth order sample and hold
(middle) and piecewise linear (bottom) interpolation.

4.1.2.2   Reconstruction from a Filtered Signal's Samples

Here we consider reconstruction of a bandlimited signal $f(t)$
from samples of

$$g(t) = f(t) * h(t)$$

taken at the Nyquist rate. Let

$$2B\,K(u) = \frac{\Pi(\frac{u}{2B})}{H(u)} \tag{4.25}$$

If $K(u)$ is bounded, we can write

$$F(u) = 2B\,G(u)\,K(u)\,\Pi(\frac{u}{2B}).$$

Using this and

$$G(u) = \frac{1}{2B} \sum_{n=-\infty}^{\infty} g(\frac{n}{2B}) e^{-j\pi nu/B} \, \Pi(\frac{u}{2B})$$

in the inversion formula gives

$$f(t) = \int_{-B}^{B} F(u) e^{j2\pi ut} du$$

$$= \sum_{n=-\infty}^{\infty} g(\frac{n}{2B}) \, k(t - \frac{n}{2B}) \qquad (4.26)$$

where

$$k(t) = \frac{1}{2B} \int_{-B}^{B} \frac{e^{j2\pi ut}}{H(u)} du \longleftrightarrow K(u).$$

As we will see in Chapter 5, (4.26) is ill posed when $K(u)$ contains a pole.

4.2   Papoulis' Generalization

There are a number of ways to generalize the manner in which data can be extracted from a signal and still maintain sufficient information to reconstruct the signal. Shannon, for example, noted that one could sample at half the Nyquist rate without information loss if, at each sample location, two sample values were taken: one of the signal and one of the signal's derivative. The details were later worked out by Linden who generalized the result to restoring from a signal sample and samples of its first $N - 1$ derivatives taken every N Nyquist intervals.

Alternately, one can choose any N distinct points within N Nyquist intervals. If signal samples are taken at these locations every N Nyquist intervals, we address the question of restoration from *interlaced* or *bunched* samples [Yen].

Another problem encountered is restoration of a signal from samples taken at half the Nyquist rate along with samples of the signal's Hilbert transform taken at the same rate.

Remarkably, all of these cases are subsumed in a generalization of the sampling theorem developed by Papoulis (1977). The generalization concerns restoration of a signal given data sampled at $1/N^{th}$ the Nyquist rate from the output of N filters into which the signal has been fed. The result is a generalization of the reconstruction from the filtered signal's samples presented in Sec.4.1.2.2.

In this section, we first present a derivation of *Papoulis' Generalized Sampling Theorem*. Specific attention is then given to interpolation function evaluation. Lastly, specific applications of the problems addressed at the beginning of this section are given.

Let $\{\ H_p(u)\ \mid\ p = 1, 2, \ldots, N\}$ be a set of N given filter frequency responses and let $f(t)$ have bandwidth B. As is shown in Fig.4.4, $f(t)$ is fed into each filter. The outputs are

$$g_p(t) = f(t) * h_p(t) \ ; \ 1 \le p \le N. \tag{4.27}$$

Each output is sampled at $1/N^{th}$ the Nyquist rate. Define

$$B_N = B/N.$$

The signal of samples obtained from the pth filter are

$$
\begin{aligned}
s_p(t) &= g_p(t)\, 2B_N\, \mathrm{comb}(2B_N t) \\
&= \sum_{n=-\infty}^{\infty} g_p(nT_N)\, \delta(t - nT_N)
\end{aligned}
\tag{4.28}
$$

where $T_N = 1/2B_N$. Our problem is to restore $f(t)$ from this set of functions or, equivalently, the sample set

$$\{g_p(nT_N) \mid 1 \le p \le N, \ -\infty < n < \infty\}.$$

We will show that

$$f(t) = \sum_{p=1}^{N} \sum_{n=-\infty}^{\infty} g_p(nT_N)\, k_p(t - nT_N) \tag{4.29}$$

where

$$k_p(t) = \int_{B-2B_N}^{B} K_p(u; t)\, e^{j2\pi u t}\, du \tag{4.30}$$

and the $K_p(u;t)$'s, if they exist, are the solutions of the simultaneous set of equations

$$2B_N \sum_{p=1}^{N} K_p(u;t) H_p(u - 2mB_N) = \exp(-j2\pi mt/T_N)$$

$$0 \leq m \leq N \quad (4.31)$$

over the parameter set $\quad 0 \leq m < N, \quad B - B_N < u < B$ and $-\infty < t < \infty$. Note that $K_p(u;t)$ and $k_p(t)$ are not Fourier transform pairs.

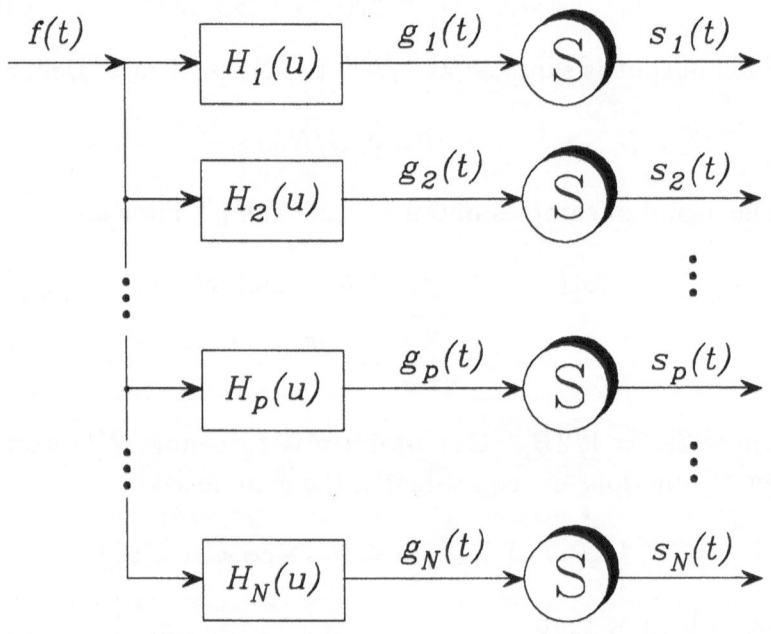

Figure 4.4: Generation of sample data from Papoulis' Generalized Sa mpling Theorem. The encircled **S** is a sampler.

4.2.1   Derivation

From (4.27)

$$G_p(u) = F(u)H_p(u).$$

Fourier transforming (4.28) and using the Fourier dual of the Poisson sum formula gives

$$S_p(u) = 2B_N \sum_{n=-\infty}^{\infty} G_p(u - 2nB_N). \qquad (4.32)$$

Clearly, each $S_p(u)$ is periodic with period $2B_N$. Since $G_p(u)$ has finite support, *i.e.*

$$G_p(u) = G_p(u)\Pi(u/2B),$$

we conclude (4.32) is simply an aliased replication of $G_p(u)$. Example replications for $N = 2$ and 3 are shown in Fig.4.5.

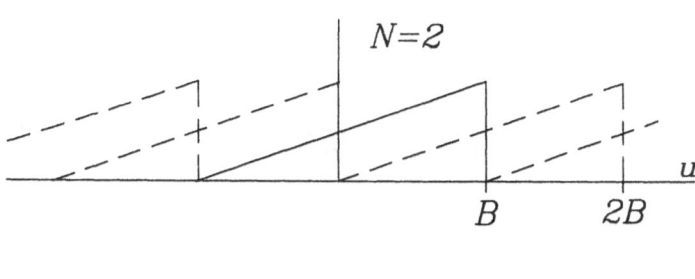

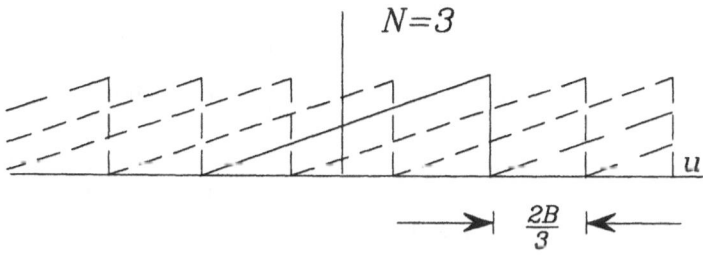

Figure 4.5: Illustration of first (top) and second order aliasing.

Note that on the interval $|u| < B$, there are $2N - 1$ portions of shifted $G_p(u)$'s. Equivalently, there are $M = N - 1$ spectra overlapping the zeroth order spectrum on both sides of the origin. Accordingly, M is referred to as the *degree of aliasing*. Over the interval

$$B - 2B_N < u < B \tag{4.33}$$

(which corresponds to one period of $S_p(u)$), there are a total of N portions of replicated spectra. If we have N varied forms of Nth order aliased data, it makes sense that our signal can be recov ered. Indeed, on the interval in (4.33),

$$S_p(u) = 2B_N \sum_{n=0}^{N-1} G_p(u - 2nB_N) \tag{4.34}$$

$$= 2B_N \sum_{n=0}^{N-1} H_p(u - 2nB_N)F(u - 2nB_N) \tag{4.35}$$

Here we have N equations and N unknowns. This may be made clearer by viewing (4.35) in matrix form:

$$2B_N \begin{bmatrix} H_1(u) & \cdots & H_1(u - 2nB_N) & \cdots & H_1(u - 2(N-1)B_N) \\ H_2(u) & \cdots & H_2(u - 2nB_N) & \cdots & H_2(u - 2(N-1)B_N) \\ \vdots & & \vdots & & \vdots \\ \vdots & & \vdots & & \vdots \\ H_p(u) & \cdots & H_p(u - 2nB_N) & \cdots & H_p(u - 2(N-1)B_N) \\ \vdots & & \vdots & & \vdots \\ \vdots & & \vdots & & \vdots \\ H_N(u) & \cdots & H_N(u - 2nB_N) & \cdots & H_N(u - 2(N-1)B_N) \end{bmatrix}$$

$$\times \begin{bmatrix} F(u) \\ F(u - 2B_N) \\ \vdots \\ \vdots \\ F(u - 2nB_N) \\ \vdots \\ \vdots \\ F(u - 2(N-1)B_N) \end{bmatrix} = \begin{bmatrix} S_1(u) \\ S_2(u) \\ \vdots \\ \vdots \\ S_p(u) \\ \vdots \\ \vdots \\ S_N(u) \end{bmatrix}$$

or, in short hand notation,

$$2B_N \mathbf{H} \vec{F} = \vec{S}. \tag{4.36}$$

Thus, assuming the $\mathbf{H}$ matrix is not singular, we can solve for $F(u)$ with knowledge of the set:

$$\{F(u - 2nB_N) \mid B - 2B_N < u < B \; ; \; 0 \leq n < N\}.$$

Indeed, each $F(u - 2nB_N)$ over the interval $B - 2B_N < u < B$ is a displaced section of $F(u)$. This is illustrated in Fig.4.6 for $N = 3$. The sections $F(u - 2nB_3)$ for $-2B_3 < u < B$ are shown there for $n = 0, 1$ and 2.

Our purpose now is to appropriately put these pieces of $F(u)$ together and inverse transform. Towards this end, let the inverse of the matrix $2B_N \mathbf{H}$ be $\mathbf{Z}$ with elements

$$\{Z_p(u; n) \mid B - 2B_N < u < B \; ; \; 1 \leq p \leq N, \; 0 \leq n < N\}.$$

That is

$$2B_N \sum_{p=1}^{N} Z_p(u; n) \qquad H_p(u - 2mB_N) = \delta[n - m];$$

$$B - 2B_N < u < B, \; 0 \leq n, m, < N.$$

The solution of (4.35) is thus

$$F(u - 2nB_N) = \sum_{p=1}^{N} S_p(u) Z_p(u; n). \tag{4.37}$$

Consider, then

$$f(t) = \int_{-B}^{B} F(\nu) e^{j2\pi\nu t} d\nu. \tag{4.38}$$

To facilitate use of (4.37), we divide the integration into N intervals of width $2B_N$:

$$\int_{-B}^{B} = \int_{B_N - 2B_N}^{B_N} + \int_{B - 4B_N}^{B - 2B_N} + \cdots + \int_{-B}^{-B + 2B_N}$$

$$= \sum_{n=0}^{N-1} \int_{B - 2(n+1)B_N}^{B - 2nB_N}$$

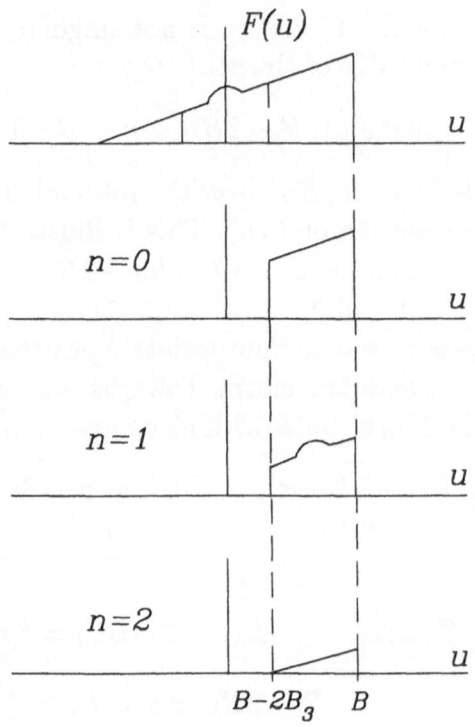

Figure 4.6: Illustration of $F(u - 2nB_N)$ on $B - 2B_N < u < B$ for various n.

Substitute into (4.38) and make the variable substitution $\nu = u - 2nB_N$.

$$f(t) = \sum_{n=0}^{N-1} \int_{B-2B_N}^{B} F(u - 2nB_N)e^{j2\pi t(u-2nB_N)}du. \qquad (4.39)$$

Define

$$K_p(u;t) = \sum_{n=0}^{N-1} Z_p(u;n) \exp(-j4\pi nB_N t). \qquad (4.40)$$

Then substitution of (4.37) into (4.39) gives

$$f(t) = \sum_{p=1}^{N} \int_{B-2B_N}^{B} S_p(u) K_p(u; t) e^{j2\pi t u} du. \qquad (4.41)$$

Directly transforming (4.28) yields

$$S_p(u) = \sum_{n=-\infty}^{\infty} g_p(nT_N) \exp(-j\pi n u / B_N).$$

Substituting into (4.41) produces our desired result:

$$f(t) = \sum_{p=1}^{N} \sum_{n=-\infty}^{\infty} g_p(nT_N) k_p(t - nT_N) \qquad (4.42)$$

where

$$k_p(t) = \int_{B-2B_N}^{B} K_p(u; t) e^{j2\pi t u} du. \qquad (4.43)$$

Equation (4.42) generates $f(t)$ from the undersampled outputs of each of the N filters.

4.2.2 Interpolation Function Computation

In order to find the k_p's required for interpolation in (4.42), for a given set of frequency responses, we need not invert the **H** matrix in (4.36). Rather, we will show that one need only solve N simultaneous equations.

To derive this set of equations, we rewrite (4.40) as

$$\vec{K} = \mathbf{Z}\,\vec{E} \qquad (4.44)$$

where the vector $\vec{K}$ has elements

$$K_p(u; t) \; ; \; B - 2B_N < u < B, \; 1 \le p \le N$$

and $\vec{E}$ has elements

$$\exp(-j4\pi n B_N t) \; ; \; -\infty < t < \infty, \; 0 \le n < N.$$

Multiplying both sides of (4.44) by $2B_N \mathbf{H} = \mathbf{Z}^{-1}$ gives

$$2B_N \mathbf{H}\vec{K} = \vec{E}$$

or, equivalently

$$2B_N \sum_{p=1}^{N} K_p(u;t)H_p(u - 2mB_N) = \exp(-j4\pi m B_N t) \quad (4.45)$$

where $0 \leq m < N$, $B - 2B_N < u < B$ and t is arbitrary. The $K_p(u;t)$'s can be determined from this set of equations and the corresponding interpolation functions from (4.43).

4.2.3   *Example Applications*

4.2.3.1   Recurrent Nonuniform Sampling

As is shown in Fig.4.7, let $\{\alpha_p \mid p = 1, 2, \ldots, N\}$ denote N distinct locations in N Nyquist intervals. A signal is sampled at these points every N Nyquist intervals. We thus have knowledge of the data

$$\{f(\alpha_p + \frac{m}{2B_N}) \mid 1 \leq p \leq N, \ -\infty < m < \infty\}.$$

Such sampling is also referred to as *bunched* or *interlaced* sampling.

The generalized sampling theorem is applicable here if we choose for filters:

$$H_p(u) = \exp(j2\pi\alpha_p u); \ 1 \leq p \leq N. \quad (4.46)$$

The corresponding equations in (4.45) can be solved in closed form, using Cramer's rule and the Vandermonde determinant [Hamming](See Exercise 4.9). On the interval $(0, T_N)$, the resulting interpolation functions are:

$$k_p(t) = \text{sinc}[2B_N(t - \alpha_p)] \prod_{\substack{q = 1 \\ q \neq p}}^{N} \frac{\sin[2\pi B_N(t - \alpha_q)]}{\sin[2\pi B_N(\alpha_p - \alpha_q)]}. \quad (4.47)$$

Note that $k_p(t)$ is a true interpolation function in the sense that

$$k_p(\alpha_n) = \delta[p - n].$$

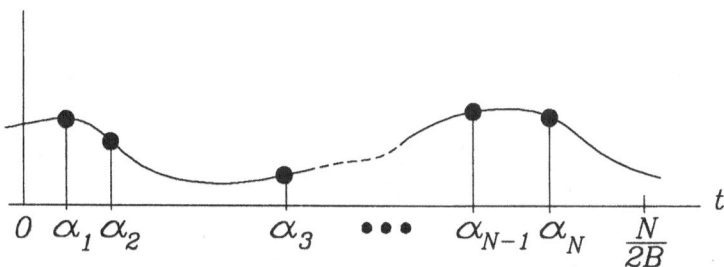

Figure 4.7: Illustration of Nth order recurrent nonuniform sampling. In each N Nyquist intervals, samples are taken at these same relative locations.

4.2.3.2   Interlaced Signal–Derivative Sampling

Consider the $N = 2$ case where

$$H_1(u) = e^{j2\pi\alpha u} \tag{4.48}$$

$$H_2(u) = (j2\pi u)^M. \tag{4.49}$$

The output of filter #1 is $f(t + \alpha)$ and that of #2 is the Mth derivative of $f(t)$. The resulting sampling geometry is shown in Fig.4.8. We will derive the spectra of the two corresponding interpolation functions.

The output of the two filters is sampled at a rate of $2B_N = B$. From (4.35), the desired signal spectrum, $F(u)$, satisfies the set of equations

$$\begin{bmatrix} S_1(u) \\ S_2(u) \end{bmatrix} = B \begin{bmatrix} e^{j2\pi\alpha u} & e^{j2\pi\alpha(u-B)} \\ (j2\pi u)^M & [j2\pi(u-B)]^M \end{bmatrix} \begin{bmatrix} F(u) \\ F(u-B) \end{bmatrix}$$

$$0 < u < B.$$

The determinant of the **H** matrix here is

$$\Delta(u) = -(j2\pi)^M e^{j2\pi\alpha u}[u^M e^{-j2\pi\alpha B} - (u-B)^M]. \tag{4.50}$$

Solving the two simultaneous equations results in

$$F(u) = \frac{\{j2\pi(u-B)\}^M S_1(u) - e^{j2\pi\alpha(u-B)} S_2(u)}{B\Delta(u)} \; ; \; 0 < u < B$$

and

$$F(u-B) = -\frac{(j2\pi u)^M S_1(u) - e^{j2\pi\alpha u} S_2(u)}{B\Delta(u)} \; ; \; 0 < u < B.$$

Shifting the second term to the interval $(-B, 0)$ and recognizing that both $S_1(u)$ and $S_2(u)$ are periodic with period B gives:

$$F(u) = K_1(u)S_1(u) + K_2(u)S_2(u),$$

where the spectra of the interpolation functions are

$$K_1(u) = \frac{(j2\pi)^M}{B}\left[-\frac{(u-B)^M}{\Delta(u)}\Pi(\frac{u}{B} - \frac{1}{2}) + \frac{(u+B)^M}{\Delta(u+B)}\Pi(\frac{u}{B} + \frac{1}{2})\right]$$

$$(4.51)$$

and

$$K_2(u) = \frac{e^{j2\pi\alpha u}}{B}\left[\frac{e^{-j2\pi\alpha B}}{\Delta(u)}\Pi(\frac{u}{B} - \frac{1}{2}) - \frac{e^{j2\pi\alpha B}}{\Delta(u+B)}\Pi(\frac{u}{B} + \frac{1}{2})\right].$$

$$(4.52)$$

We will use these results in Chapter 5 to show that interpolation here becomes unstable (ill–posed) when

(a) M is even and $\alpha = 0$, or

(b) M is odd and $\alpha = \frac{1}{2B}$.

Otherwise, interpolation can be tolerant of data noise.

Reconstruction from the $M = 1, \alpha = 0$ data was first addressed by Shannon and derived by Linden. Inverse transforming (4.51) and (4.52) for this case gives the interpolation functions

$$k_1(t) = \text{sinc}^2(Bt) \tag{4.53}$$

and

$$k_2(t) = t\,\text{sinc}^2(Bt) \tag{4.54}$$

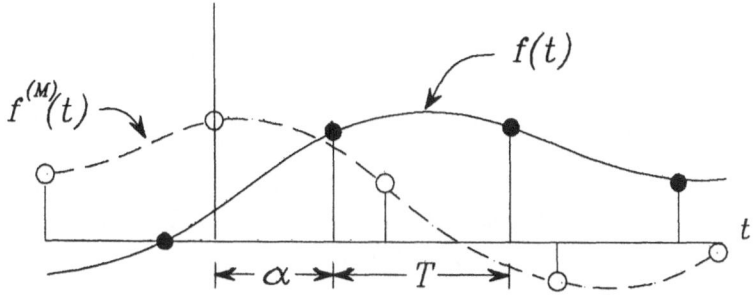

Figure 4.8: Interlaced signal – derivative sampling. The hollow
dots represent samples of the Mth derivative and the
solid dots are signal samples.

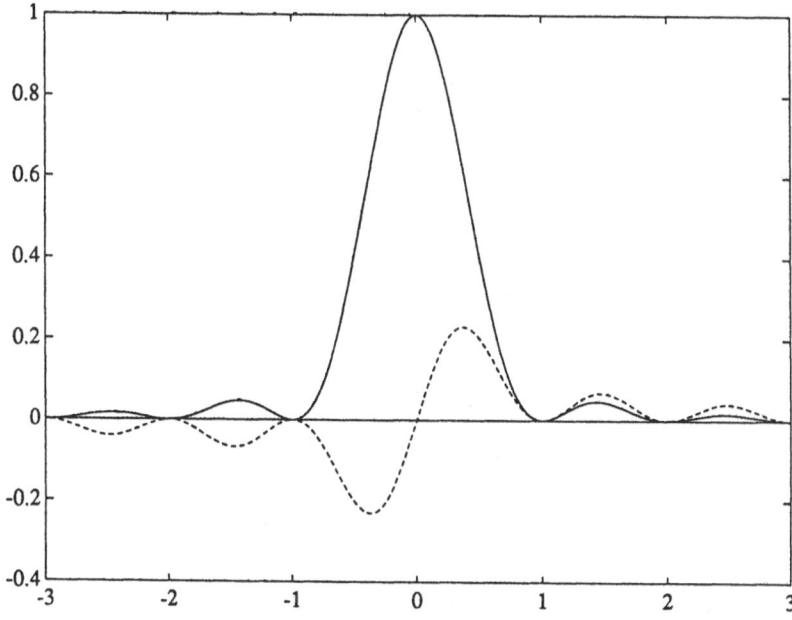

Figure 4.9: The functions for interpolating a signal from its samples and samples of its derivatives each taken simultaneously at half the Nyquist rate. The function $k_1(t/B)$,
(shown with the solid line), is used for the samples
and $k_2(t/B)$, (broken line), for the derivative samples.

These are pictured in Fig. 4.9. It follows that

$$f(t) = \frac{\sin^2(\pi Bt)}{\pi^2} \sum_{n=-\infty}^{\infty} \left[\frac{f(\frac{n}{B})}{(Bt-n)^2} + \frac{f'(\frac{n}{B})}{B(Bt-n)} \right]. \qquad (4.55)$$

4.2.3.3   Higher Order Derivative Sampling

Consider sampling a signal and its first $N-1$ derivatives every N Nyquist intervals [Linden & Abramson]. We can show that, as $N \to \infty$, the interpolation functions for the restoration approach those used in a Taylor series expansion.

The filters required for our problem are

$$H_p(u) = (j2\pi u)^{p-1} \; ; 1 \le p \le N. \qquad (4.56)$$

The solution for the interpolation function for the $N = 1$ case is clearly

$$k_1(t) = \text{sinc}(2Bt) \; ; \quad N = 1.$$

For $N = 2$, the interpolation functions are given by (4.53) and (4.54) which we rewrite here as

$$k_1(t) = \text{sinc}^2(\frac{2Bt}{2})$$
$$k_2(t) = t\,\text{sinc}^2(\frac{2Bt}{2}).$$

In Exercise 4.23, we show that, for $N = 3$

$$k_1(t) = \text{sinc}^3(\frac{2Bt}{3})$$
$$k_2(t) = t\,\text{sinc}^3(\frac{2Bt}{3}) \qquad (4.57)$$
$$k_3(t) = \frac{1}{2}t^2\,\text{sinc}^3(\frac{2Bt}{3}).$$

From this pattern, we deduce that, in general,

$$k_p(t) = \frac{t^{p-1}\,\text{sinc}^N(\frac{2Bt}{N})}{(p-1)!} \; ; \; 1 \le p \le N.$$

Substituting into (4.42) gives the interpolation series

$$f(t) = \sum_{p=1}^{N} \sum_{n=-\infty}^{\infty} \frac{\left(t - \frac{nN}{2B}\right)^{p-1}}{(p-1)!} f^{(p-1)}\left(\frac{nN}{2B}\right) \operatorname{sinc}^{N}\left(\frac{2Bt}{N} - n\right).$$

Since

$$\lim_{N \to \infty} \operatorname{sinc}^{N}\left(\frac{2Bt}{N} - n\right) = \delta[n],$$

we conclude that

$$\lim_{N \to \infty} f(t) = \sum_{p=1}^{\infty} \frac{t^{p-1}}{(p-1)!} f^{(p-1)}(0)$$

which is recognized as the Taylor series expansion of $f(t)$ about $t = 0$.

4.2.3.4   Effects of Oversampling

Suppose $f(t)$ has bandwidth B and is sampled at a rate of $2W > 2B$. Redefine $T_N = N/2W$. Then the transform of (4.42) becomes

$$F(u) = \sum_{p=1}^{N} K_p(u) \sum_{n=-\infty}^{\infty} g_p(nT_N)e^{-j2\pi nuT_N}\Pi\left(\frac{u}{2W}\right).$$

Multiplying both sides by $\Pi(u/2B)$ leaves the result unaltered:

$$F(u) = \sum_{p=1}^{N} K_p(u) \sum_{n=-\infty}^{\infty} g_p(nT_N)e^{-j2\pi nuT_N}\Pi\left(\frac{u}{2B}\right). \qquad (4.58)$$

In the time domain, this is equivalent to using the interpolation function set $\{\hat{k}_p(t)\}$ in place of $\{k_p(t)\}$ where

$$\begin{aligned}
\hat{k}_p(t) &= \int_{-B}^{B} K_p(u)e^{j2\pi ut}du \\
&= k_p(t) * 2B \operatorname{sinc}(2Bt). \qquad (4.59)
\end{aligned}$$

Inverse transforming (4.58) then gives us the oversampled version of (4.42):

$$f(t) = \sum_{p=1}^{N} \sum_{n=-\infty}^{\infty} g_p(nT_N)\hat{k}_p(t - nT_N). \qquad (4.60)$$

As we shall demonstrate in Chapter 5, oversampling can reduce interpolation noise level due to noisy data. Thus, with all other factors equal, (4.60) should be used in lieu of (4.42) for interpolating oversampled signals.

4.3   Derivative Interpolation

Interpolation formulae for generating the derivative of a bandlimited signal can be obtained by direct differentiation of the cardinal series [Marks and Hall]. The result is

$$
\begin{aligned}
x^{(p)}(t) &= (\frac{d}{dt})^p x(t) \\
&= (2B)^p \sum_{n=-\infty}^{\infty} x(\frac{n}{2B}) d_p(2Bt - n) \quad (4.61)
\end{aligned}
$$

where

$$
d_p(t) = (\frac{d}{dt})^p \operatorname{sinc}(t)
$$

is the derivative kernel. From the derivative theorem of Fourier analysis, we can equivalently write

$$
\begin{aligned}
d_p(t) &= \int_{-1/2}^{1/2} (j2\pi u)^p e^{j2\pi ut} du \\
&= \frac{(-1)^p p!}{\pi t^{p+1}} [\sin(\pi t) \cos_{p/2}(\pi t) - \cos(\pi t) \sin_{\frac{p-1}{2}}(\pi t)]
\end{aligned}
$$
$$(4.62)$$

where the incomplete sine and cosine are defined, respectively, as

$$
\cos_a(z) = \sum_{n=0}^{[a]} \frac{(-1)^n z^{2n}}{(2n)!} \quad (4.63)
$$

and

$$
\sin_a(z) = \sum_{n=0}^{[a]} \frac{(-1)^n z^{2n+1}}{(2n+1)!}. \quad (4.64)
$$

The notation [a] denotes "the greatest integer less than or equal to a". To allow for $p = 0$ in (4.62), we set $\sin_{-1/2}(t) = 0$. Then

$d_0(t) = \text{sinc}(t)$. In the evaluation of (4.62), we used the identity [Gradshteyn and Ryzhik]

$$\int x^n \, e^{ax} dx = (-1)^n \, e^{ax} \sum_{k=0}^{n} (-1)^k \, \frac{(n-k)!}{k!} \, \frac{x^k}{a^{n-k+1}}.$$

4.3.1   Properties of the Derivative Kernel

This section will be devoted to exploring properties of the derivative kernel. For large t and even p, the $\cos_{p/2}(\pi t)$ term in (4.62) dominates. For odd p, $\sin_{(p-1)/2}(\pi t)$ dominates. This observation leads to the following asymptotic relation for $d_p(t)$ for large t:

$$\lim_{t \to \infty} d_p(t) = \begin{cases} (-1)^{p/2} \pi^p \, \text{sinc}(t) & ; \quad p \text{ even} \\ (-1)^{(p-1)/2} \pi^p \cos(\pi t)/(\pi t) & ; \quad p \text{ odd}. \end{cases}$$

Convolution of $(2B)^{p+1} d_p(2Bt)$ with any bandlimited $x(t)$ yields $x^{(p)}(t)$. To show this, we write

$$(2B)^{p+1} \int_{-\infty}^{\infty} x(\tau) d_p[2B(t-\tau)] d\tau$$

$$= (-1)^p \int_{-\infty}^{\infty} x(\tau) \left(\frac{d}{d\tau} \right)^p \text{sinc}[2B(t-\tau)] d\tau$$

$$= \int_{-B}^{B} X(u)(j2\pi u)^p \, e^{j2\pi ut} du$$

$$= x^{(p)}(t) \qquad\qquad (4.65)$$

where, in the second step, we have used the power theorem of Fourier analysis. This result is a generalization of that of Gallagher and Wise who noted that the first derivative of a bandlimited signal can be achieved by a convolution with an appropriately scaled first–order spherical Bessel function $j_1(t) = -d/dt \, \text{sinc}(t/\pi)$.

Using $d_q(t)$ as the signal in (4.65) gives the recurrence relation

$$d_{p+q}(t) = \int_{-\infty}^{\infty} d_q(\tau) d_p(t-\tau) d\tau. \qquad\qquad (4.66)$$

Thus, higher order kernels can be generated by convolution of lower ordered kernels.

A second obvious recurrence relation is

$$d_{p+q}(t) = (\frac{d}{dt})^p \, d_q(t).$$

Using this expression with $q = 1$ and the relations

$$\frac{d}{dt} \cos_n(t) = -\sin_{n-1}(t)$$

$$\frac{d}{dt} \sin_n(t) = \cos_n(t)$$

gives, via (4.62), a third recurrence formula:

$$\frac{d}{dt} d_p(t) = d_{p+1}(t)$$

$$= \begin{cases} \frac{-(p+1)}{t} d_p(t) + \frac{(-1)^{p/2} \pi^p}{t} \cos(\pi t); & \text{even } p \\ \frac{-(p+1)}{t} d_p(t) - \frac{(-1)^{\frac{p-1}{2}} \pi^p}{t} \sin(\pi t); & \text{odd } p. \end{cases}$$

$$(4.67)$$

Alternate derivative interpolation can be achieved by recognizing that if $x(t)$ is bandlimited, so is $x^{(p)}(t)$. Therefore

$$x^{(p)}(t) = \sum_{m=-\infty}^{\infty} x^{(p)}(\frac{m}{2B}) \, \text{sinc}(2Bt - m).$$

Thus, the signal derivative is uniquely specified by its sample values which , from (4.61) can be computed by the discrete convolution

$$x^{(p)}(\frac{m}{2B}) = (2B)^p \sum_{n=-\infty}^{\infty} x^{(p)}(\frac{n}{2B}) \, d_p(m - n) \qquad (4.68)$$

where, from (4.62)[1]

$$d_p(m) = \begin{cases} \frac{-(-1)^{m+p} p!}{\pi m^{p+1}} \sin_{\frac{p-1}{2}}(\pi m) & ; \quad m \neq 0 \\ (-1)^p \frac{\pi^p}{p+1} \delta[p - \text{even}] & ; \quad m = 0. \end{cases}$$

Note that the discrete derivative kernel is independent of the signal bandwidth. Plots of $|d_p(m)|$ are shown in Figs. 4.10 and 4.11.

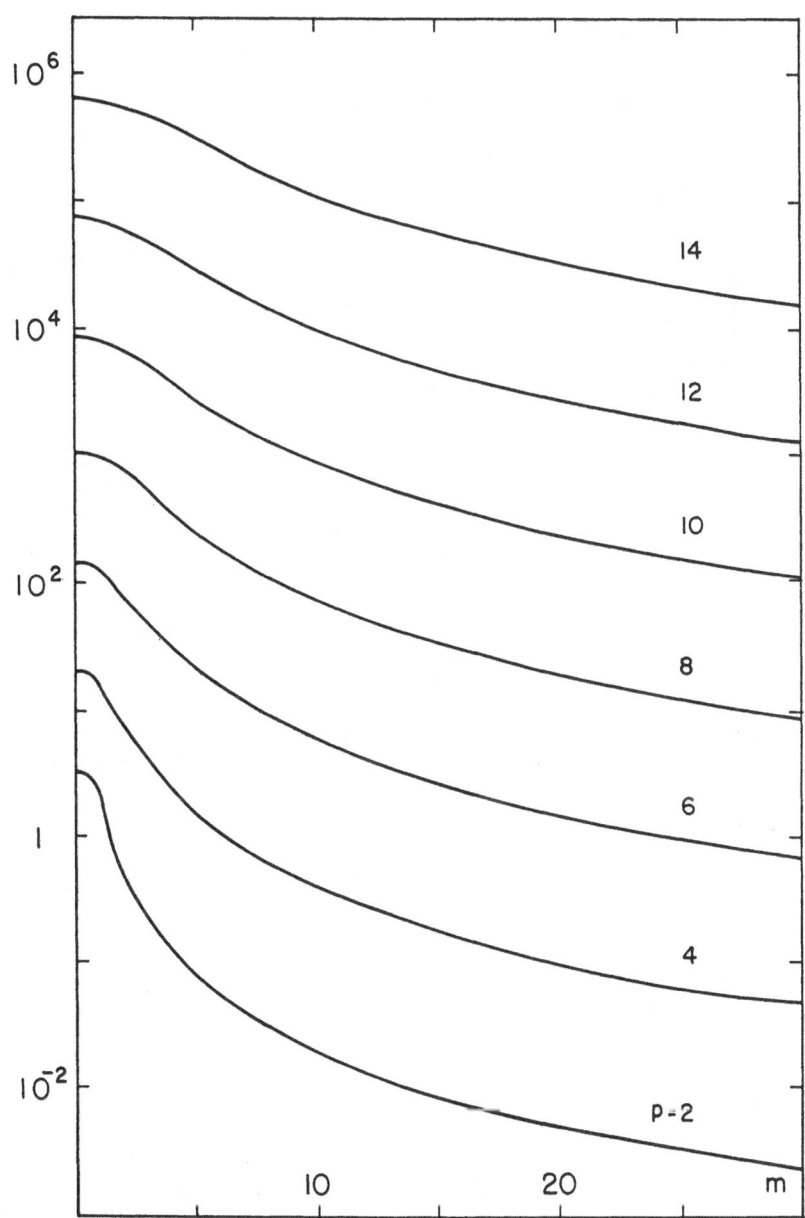

Figure 4.10: Plots of $|d_p(m)|$ for even p. Points are connected for clarity.

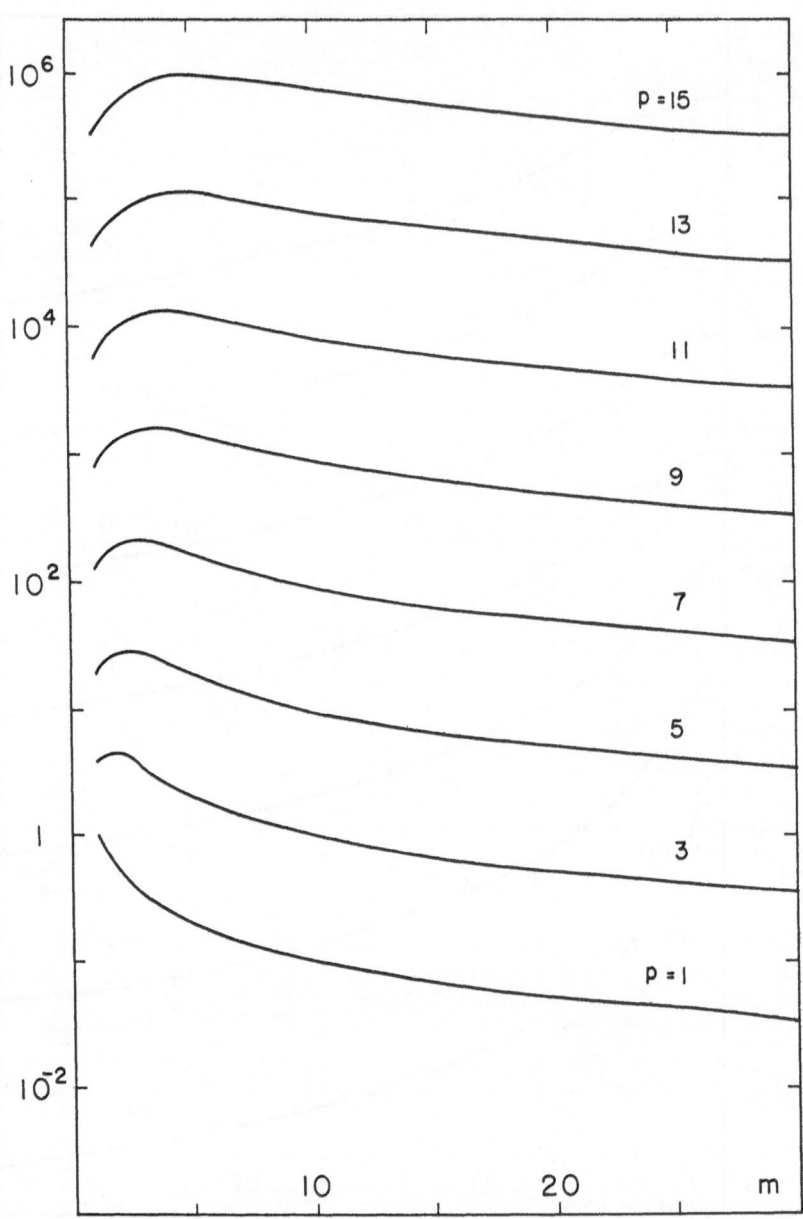

Figure 4.11: Plots of $|d_p(m)|$ for odd p.

Using (4.64), the asymptotic behavior for $d_p(m)$ for large m is found to be

$$d_p(m) \longrightarrow \begin{cases} (-1)^{m+\frac{p}{2}}p\pi^{p-2}/m^2 & ; \quad \text{even } p \\ \frac{(-1)^{m+\frac{p-1}{2}}\pi^{p-1}}{m} & ; \quad \text{odd } p. \end{cases}$$

A recurrence relation for the discrete derivative kernel follows from the use of $d_q(n)$ as the signal in (4.68)

$$d_{p+q}(m) = \sum_{n=-\infty}^{\infty} d_q(n)d_p(m-n).$$

This is the discrete equivalent of (4.66).

A second recurrence relation immediately follows from (4.67) for $m \neq 0$:

$$d_{p+1}(m) = \begin{cases} \frac{-(p+1)}{m}d_p(m) + \frac{(-1)^{m+\frac{p}{2}}\pi^p}{m} & ; \quad \text{even } p \\ \frac{-(p+1)}{m}d_p(m) & ; \quad \text{odd } p. \end{cases}$$

The discrete derivative kernel is square summable. Since $d_p(m)$ is simply the mth Fourier coefficient of $(j2\pi u)^m$ for $|u| < 1/2$, we have

$$\sum_{m=-\infty}^{\infty} |d_p(m)|^2 = \int_{-\infty}^{\infty} |d_p(t)|^2 dt$$

$$= \int_{-1/2}^{1/2} |(j2\pi u)^p|^2 du$$

$$= \frac{\pi^{2p}}{2p+1}.$$

The sensitivity of derivative interpolation to additive sample noise is examined in Chapter 5. There, we show that the interpolation noise level increases significantly with p.

---

[1]To derive the $m = 0$ case, it is easiest to use the integral in (4.66) with $t = m = 0$.

4.4   A Relation Between the Taylor and Cardinal Series

The discrete derivative kernel can be utilized to couple a band-limited signal's Taylor series and sampling theorem expansion. If $x(t)$ is bandlimited, it is analytic everywhere. Thus, its Taylor series about $m/2B$:

$$x(t) = \sum_{p=0}^{\infty} \frac{(t - \frac{m}{2B})^p}{p!} x^{(p)} \left(\frac{m}{2B}\right),$$

converges for all t. Substituting (4.68) gives

$$x(t) = \sum_{p=0}^{\infty} \frac{(2Bt - m)^p}{p!} \sum_{n=-\infty}^{\infty} x(\frac{n}{2B}) \, d_p(m - n).$$

Since the series is absolutely convergent (see Exercise 2.22), we can interchange the summation order:

$$x(t) = \sum_{n=-\infty}^{\infty} x(\frac{n}{2B}) \sum_{p=0}^{\infty} \frac{(2Bt - m)^p d_p(m - n)}{p!}. \qquad (4.69)$$

The sum over p is recognized as the Taylor series expansion of $\mathrm{sinc}(2Bt - n)$ about $t = m/2B$. Thus (4.69) reduces to the cardinal series.

4.5   Sampling Trigonometric Polynomials

A trigonometric polynomial is a bandlimited periodic function with a finite number of nonzero Fourier coefficients. A low pass trigonometric polynomial with period T can be written as

$$x(t) = \sum_{m=-N}^{N} c_m e^{-j2\pi mt/T}. \qquad (4.70)$$

This function is uniquely determined by $2N + 1$ coefficients. We therefore would expect that $2N+1$ samples taken within a single period would suffice to uniquely specify $x(t)$. We will show that,

$$x(t) = \sum_{q=1}^{P} x(qT_p) \, k(\frac{t}{T_p} - q) \qquad (4.71)$$

where $T_p = T/P$ is the sampling interval, P (assumed odd) is the number of samples per period and the interpolation function is

$$k(t) = \frac{\sin(\pi t)/P}{\sin(\pi t/P)}.\tag{4.72}$$

We require that $P > 2N + 1$. Plots of $k(t)$ for various P are shown in Fig.4.12.

Proof : The cardinal series for $x(t)$ can be written as

$$x(t) = \sum_{p=-\infty}^{\infty} x(pT_p) \operatorname{sinc}\left(\frac{t}{T_p} - p\right)\tag{4.73}$$

where the sampling interval is

$$T_p = \frac{T}{P} \; ; \quad P = 2M + 1 > 2N + 1.\tag{4.74}$$

We have assumed, for simplicity, that the odd number of samples taken in each period are the same. We can partition the sum in (4.73) as

$$\sum_{p=-\infty}^{\infty} = \ldots + \sum_{p=1+P}^{2P} + \sum_{p=1}^{P} + \sum_{p=1-P}^{0} + \ldots$$

$$= \sum_{n=-\infty}^{\infty} \sum_{p=1-nP}^{(1-n)P}$$

$$= \sum_{n=-\infty}^{\infty} \sum_{q=1}^{P}$$

where $q = p + nP$. Using this, and recognizing that $x[(q - nP)T_p] = x(qT_p)$ reduces (4.73) to

$$x(t) = \sum_{q=1}^{P} x(qT_p) \, i_q(t)$$

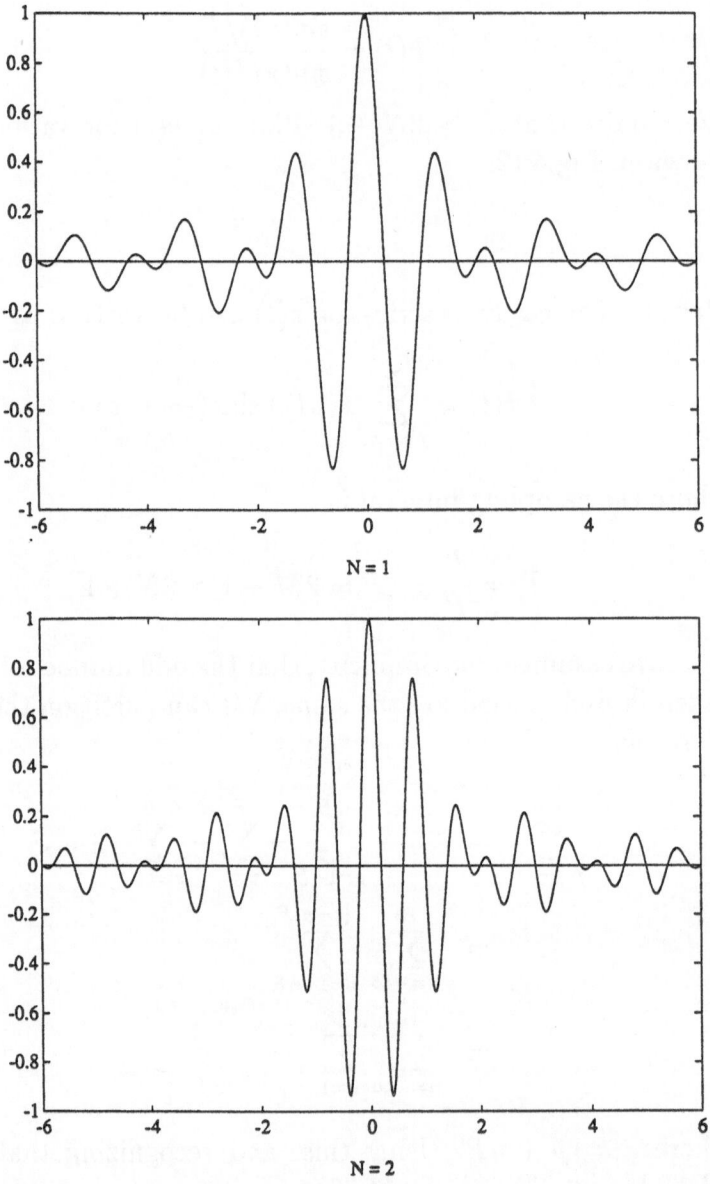

Figure 4.12: Plots of the interpolation functions for trigonometric polynomials. The number of samples taken per period is P.

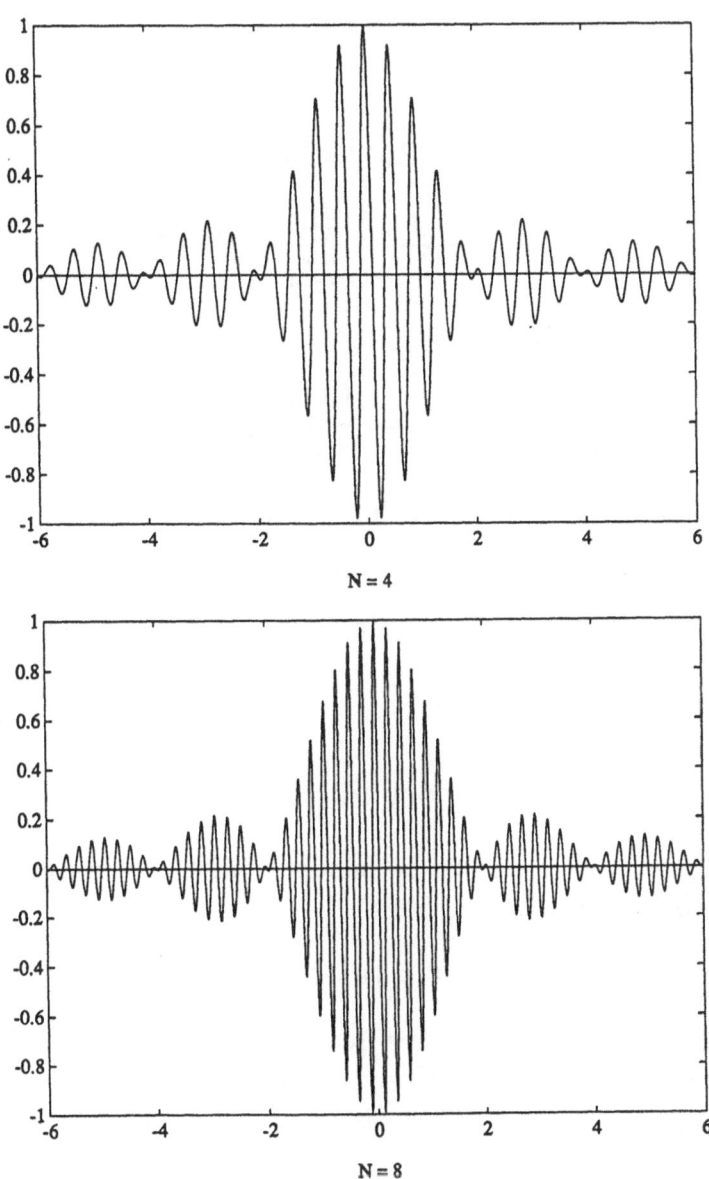

N = 4

N = 8

Continuation of Fig 4.12.

where the interpolation function is

$$i_q(t) = \sum_{n=-\infty}^{\infty} \text{sinc}[\frac{t+nT}{T_p} - q].$$ (4.75)

Using the Poisson sum formula

$$i_q(t) = \frac{1}{P} \sum_{n=-\infty}^{\infty} \Pi(\frac{n}{P}) \, e^{-j2\pi qn/P} e^{j2\pi nt/T}$$

$$= \frac{1}{2M+1} \sum_{n=-M}^{M} e^{-j2\pi n(\frac{q}{P} - \frac{t}{T})}$$

and the geometric series

$$\sum_{m=-M}^{M} z^m = \frac{z^{(M+\frac{1}{2})} - z^{-(M+\frac{1}{2})}}{z^{1/2} - z^{-1/2}}$$

gives, after application of Euler's formula

$$i_q(t) = \frac{\sin[\pi(\frac{t}{T_p} - q)]/P}{\sin[\pi(\frac{t}{T_p} - q)/P]}$$

and our proof is complete.

4.6 Sampling Theory for Bandpass Functions

A signal $x(t)$ is said to be bandpass with center frequency f_0 and bandwidth B if

$$X(u) \equiv 0 ; \quad 0 < |u| < f_L, \; f_U < |u| < \infty$$

where the upper and lower frequencies are $f_L = f_0 - B/2$ and $f_U = f_0 + B/2$ respectively. An example spectrum is shown in Fig. 4.13. The signal is assumed to be real so that $X(u)$ is Hermetian.

We will discuss two techniques to characterize a bandpass function by its samples. The first requires preprocessing prior to sampling. The second uses samples taken directly from $x(t)$ at a rate of $2B$. A hybrid approach is left as an exercise.

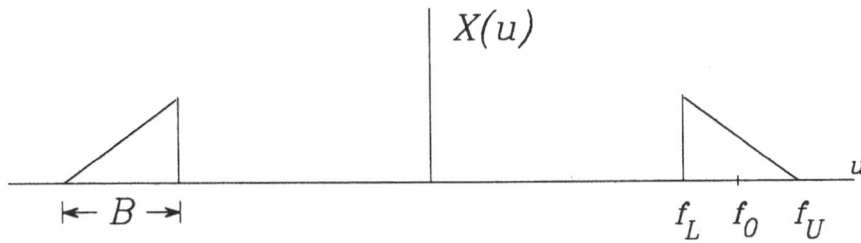

Figure 4.13: The spectrum of a bandpass signal of center frequency f_0 and bandwidth B.

4.6.1 Heterodyned Sampling

A bandpass signal can be heterodyned down to baseband by using the standard (coherent) upper sideband amplitude demodulation technique illustrated in Fig. 4.14. The bandpass signal is first multiplied by a cosinusoid to obtain

$$y(t) = x(t) \cos(2\pi f_L t)$$

or, in the frequency domain

$$
\begin{aligned}
Y(u) &= X(u) * \frac{1}{2} [\delta(u - f_L) + \delta(u + f_L)] \\
&= \frac{1}{2} X(u - f_L) + \frac{1}{2} X(u + f_L).
\end{aligned}
$$

The result is illustrated in Fig. 4.15. The signal $y(t)$ is then low pass filtered to yield the baseband signal $z(t)$, which can be sampled by conventional means. If sampling is performed at the Nyquist rate, then the baseband samples are

$$z(\frac{n}{2B}) = 2B \int_{-\infty}^{\infty} x(t) \cos(2\pi f_L t) \operatorname{sinc}(2Bt - n) dt.$$

Postprocessing is also required to regenerate $x(t)$ from samples of $z(t)$. We partition $Z(u)$ as

$$Z(u) = U(u) + L(u)$$

where

$$U(u) = Z(u)\,\mu(u)$$

and

$$L(u) = Z(u)\,\mu(-u).$$

Then, clearly

$$\begin{aligned}\frac{1}{2}X(u) &= L(u + f_L) + U(u - f_L)\\ &= Z(u + f_L)\mu(-u - f_L) + Z(u - f_L)\,\mu(u - f_L).\end{aligned}$$

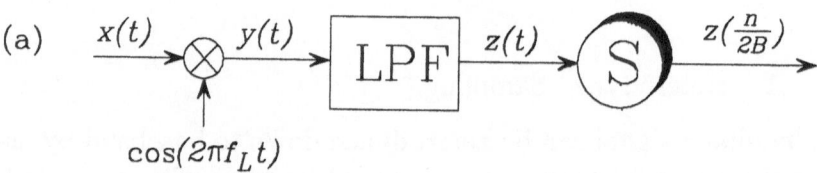

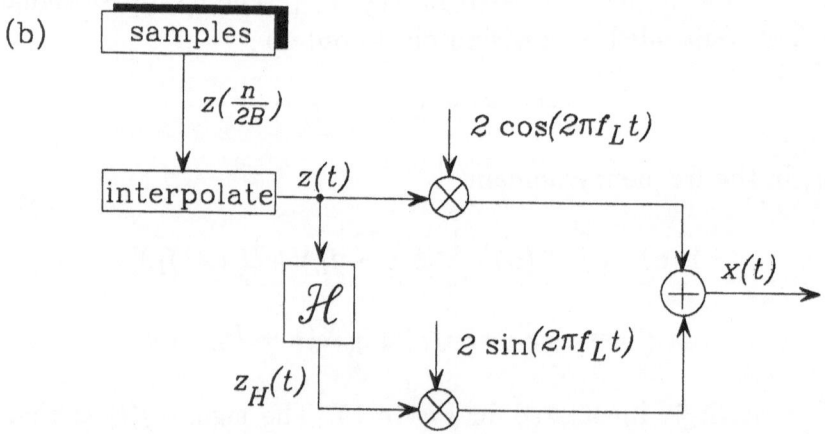

Figure 4.14: (a) Heterodyning a bandpass signal to baseband in order to apply conventional sampling. The encircled S is a sampler. (b) Restoration of the bandpass signal from the baseband signal's samples.

The inverse transform of the first term is the conjugate of the inverse transform of the second. Thus

$$\frac{1}{2}x(t) = 2\Re\, w(t)$$

where

$$w(t) = \{z(t)\exp(-j2\pi f_L t)\} * \{[\frac{1}{2}\delta(t) - \frac{j}{2\pi t}]\exp(-j2\pi f_L t)\},$$

and $\Re$ is the real operator. Simplifying gives

$$x(t) = 2\,z(t)\cos(2\pi f_L t) + 2\,z_H(t)\sin(2\pi f_L t)$$

where the *Hilbert transform* of $z(t)$ is

$$z_H(t) = \frac{-1}{\pi}\int_{-\infty}^{\infty}\frac{z(\tau)\,d\tau}{t-\tau}.$$

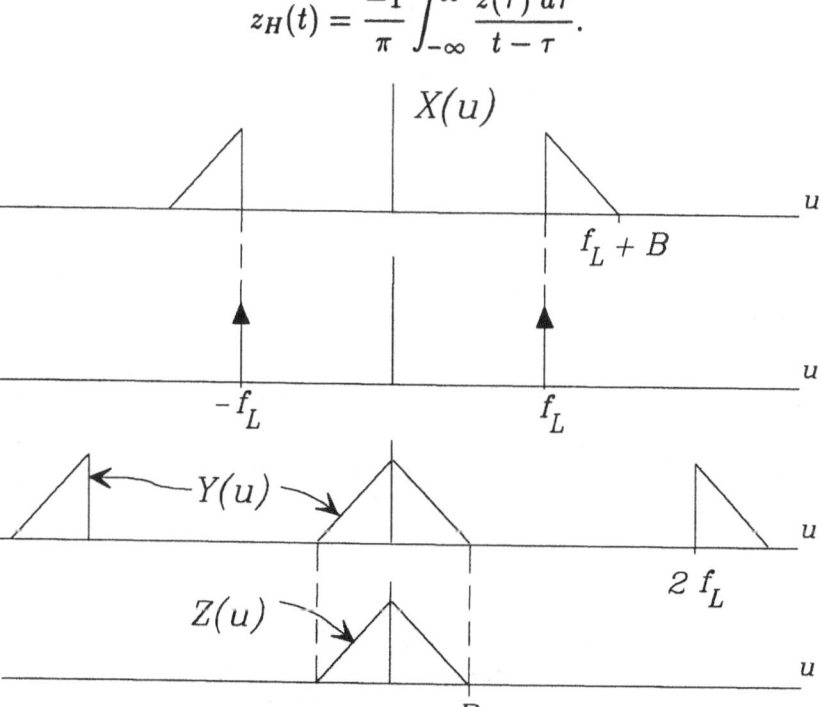

Figure 4.15: Illustration of heterodyning down to baseband.

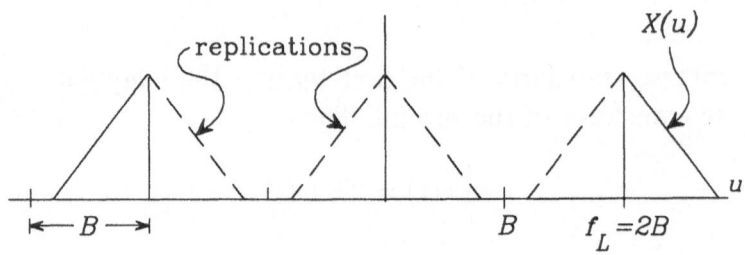

Figure 4.16: When f_L is an integer multiple of B, the spectral replications corresponding to a sampling rate of $2B$ do not overlap the original spectrum. Note, in this example, we had to artificially increase B to meet the integer multiplication criterion.

4.6.2   Direct Bandpass Sampling

A bandpass signal can also be reconstructed by samples taken directly from the signal. With reference to Fig. 4.13, assume that f_L is an integer multiple of B:

$$f_L = 2NB. \qquad (4.76)$$

This relation can always be achieved by artificially increasing f_U, resulting in an equal incremental increase in B.

The reason for requiring (4.76) is made evident in Fig. 4.16. When the bandpass signal is sampled at a rate of $2B$, the replicated spectra do not overlap with $X(u)$ (shown with solid lines). Therefore, $X(u)$ can be regained from the sample data with the use of a bandpass filter.

Let's derive the specifics. Let

$$
\begin{aligned}
s(t) &= \sum_{n=-\infty}^{\infty} x(\frac{n}{2B}) \, \delta(t - \frac{n}{2B}) \\
&= x(t) \, 2B \, \text{comb}(2Bt) \qquad (4.77)
\end{aligned}
$$

so that

$$S(u) = X(u) * \text{comb}(u/2B)$$
$$= 2B \sum_{n=-\infty}^{\infty} X(u - 2nB).$$

The signal is regained with a bandpass filter:

$$X(u) = \frac{1}{2B} S(u) \left[\Pi(\frac{u + f_0}{B}) + \Pi(\frac{u - f_0}{B}) \right].$$

Inverse transforming gives

$$x(t) = s(t) * [\text{sinc}(Bt) \cos(2\pi f_0 t)].$$

Substituting (4.77) and simplifying leaves

$$x(t) = \sum_{n=-\infty}^{\infty} x(\frac{n}{2B}) k(2Bt - n) \qquad (4.78)$$

where the interpolation function is

$$k(t) = \text{sinc}(\frac{t}{2}) \cos[\pi(2N + 1)t/2]. \qquad (4.79)$$

Plots of $k(t)$ for various N are shown in Fig. 4.17.

4.7   A Summary of Sampling Theorems for Directly Sampled Signals

A number of the sampling theorems discussed in this chapter can be written as

$$x(t) = \sum_{n \in S} x(t_n) k_n(t) \qquad (4.80)$$

where S is a set of integers. A list of some applicable sampling theorems are in Table 4.1. The signal $x(t)$ is assumed to be (low pass or high pass) bandlimited with bandwidth B. The sampling rate $2W$ exceeds the Nyquist rate. Note that, in each

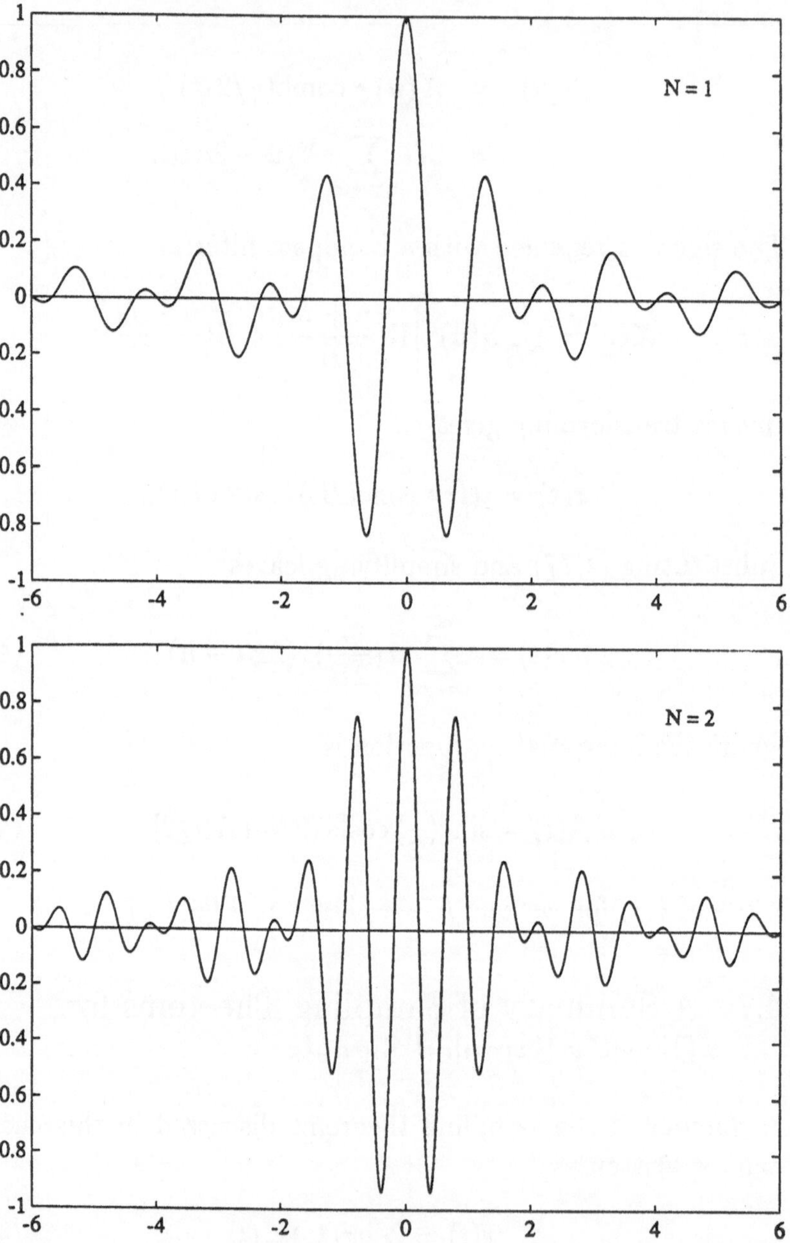

Figure 4.17: Interpolation functions for direct bandpass signal sampling for various N.

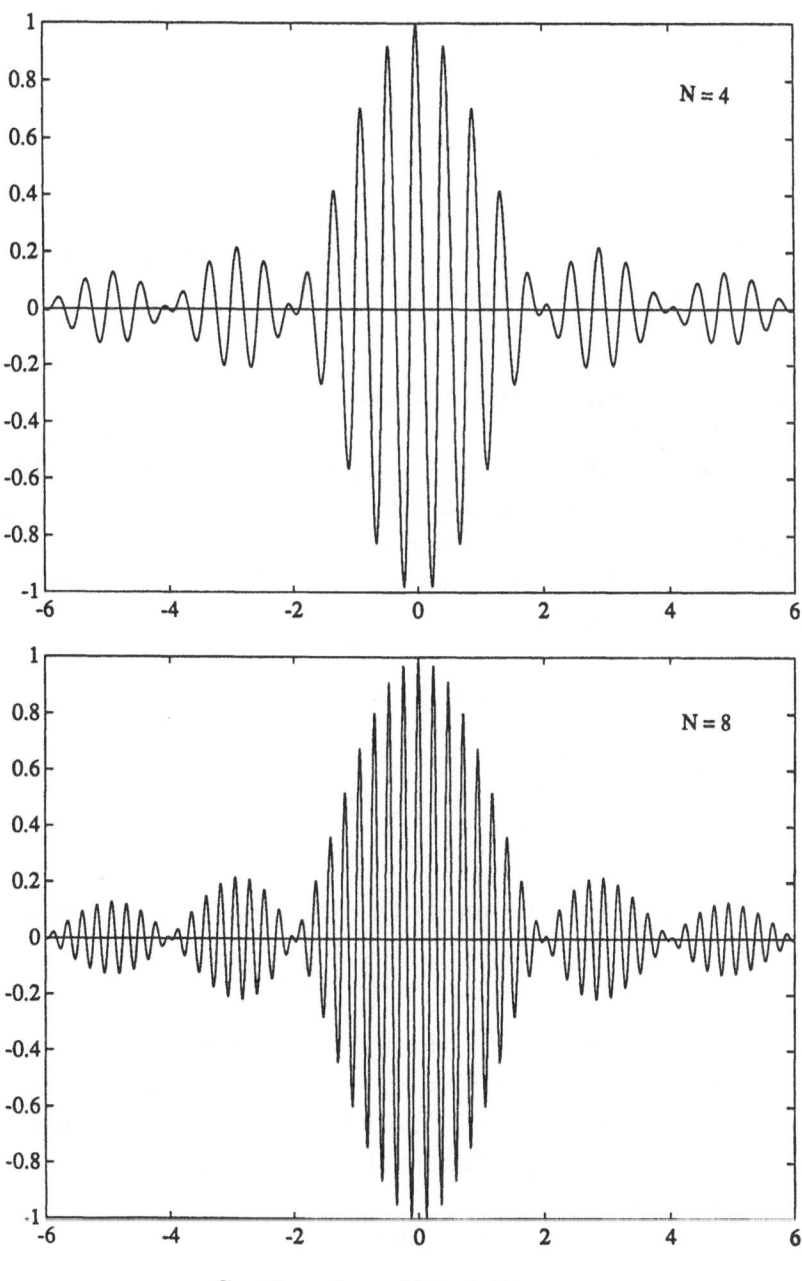

Continuation of Fig 4.17.

case, the function used for interpolation can itself be sampled and interpolated as

$$k_m(t) = \sum_{n \in S} k_m(t_n)\, k_n(t).$$

(4.81)

Excluded from this generalization are interpolations requiring samples of a signal's derivative or Hilbert transform. Derivative interpolation is likewise not included.

4.8   Lagrangian Interpolation

Lagrangian interpolation, when applied to uniformly spaced samples, is equivalent to cardinal series interpolation. In general, let $\{t_n\}$ denote a set of sample locations for a function $x(t)$. The corresponding Lagrangian interpolation from these samples is [Ralston and Rabinowitz]

$$y(t) = \sum_n x(t_n) k_n(t)$$

(4.82)

where

$$k_n(t) = \prod_{m \neq n} \frac{t - t_m}{t_n - t_m}.$$

Note that this function meets the interpolation function criterion

$$k_n(t_m) = \delta[n - m].$$

In other words, $y(t)$ passes through all the sample points.

We now show that if $\{t_n = nT | -\infty < n < \infty\}$, then (4.82) becomes the cardinal series. Under this assumption, the interpolation function clearly takes on the same form at every sample location. Thus

$$k_n(t) = k(t - nT).$$

Analysis of the $n = 0$ case therefore suffices. There,

$$k(t) = \prod_{m \neq 0} \left[1 - \left(\frac{t}{mT}\right)\right].$$

	S	t_n	$K_n(t)$	equation
Cardinal series	$-\infty < n < \infty$	$\frac{n}{2B}$	$\text{sinc}(2Bt - n)$	3.3
Oversampling	$-\infty < n < \infty$	$\frac{n}{2W}$	$r\,\text{sinc}(2Bt - rn)$	4.2
Lost sample at $t = 0$	$n \neq 0$	$\frac{n}{2W}$	$r\,\text{sinc}(2Bt - rn) + \frac{r^2}{1-r}\text{sinc}(rn)\text{sinc}(2Bt)$	4.86
Recurrent nonuniform sampling*	$1 \leq p \leq N$ $-\infty < n < \infty$	$t_{np} = \frac{n}{2B_N} + \alpha_p$	$\text{sinc}\,2B_N(t - t_{np})$ $\prod_{q=1,q\neq p}^{N} \frac{\sin[2\pi B_N(t-t_{nq})]}{\sin[2\pi B_N(t_{np}-t_{nq})]}$	4.47
Bandpass signals	$-\infty < n < \infty$	$\frac{n}{2B}$	$\text{sinc}(Bt - \frac{n}{2})$ $\cos[\pi(2N+1)(Bt - \frac{n}{2})]$	4.78 & 4.79
Trigonometric polynomials†	$1 \leq n \leq P$	$\frac{n}{2W}$	$\dfrac{\sin[\pi(2Wt - n)]/P}{\sin[\pi(2Wt - n)/P]}$	4.71 & 4.72

Table 4.1: Direct sample interpolation following the formula in (4.80).

PARAMETERS: $B =$ signal's bandwidth $W > B$

$r = B/W =$ sampling rate parameter

$B_N = B/N$

$\alpha_p =$ sample locations in recurrent nonuniform sampling

FOOTNOTES: * here, the sum is over both n and p

† $2W = P/T$ where T is the signal's period & P is an odd integer.

Separating the product into its positive and negative m portions followed by a multiplicative combination gives

$$k(t) = \prod_{m=1}^{\infty} \left[1 - \left(\frac{t}{mT} \right)^2 \right].$$

Since [Abramowitz and Stegun]

$$\sin(z) = z \prod_{m=1}^{\infty} \left[1 - \left(\frac{z}{\pi m} \right)^2 \right]$$

we conclude that

$$k(t) = \text{sinc} \left(\frac{t}{T} \right)$$

and our equivalence demonstration is complete.

4.9   Kramer's Generalization

The generalization of the sampling theorem by Kramer (1959) can best be explained by a review of the sampling theorem derivation in Section 3.2.2. followed by a parallel generalized derivation.

Consider the inverse Fourier transform expression of a band-limited function in (3.2). We can evaluate this expression without loss of information at the points $t = n/2B$ because the functions $\{\exp(-j\pi nu/B)| - \infty < n < \infty\}$ form a complete orthogonal basis set on the interval $-B < u < B$ [Luenberger; Naylor & Sell]. Therefore, as explained in Section 2.2.1.2, the inner products expressed in (3.9) are sufficient for an orthogonal series expansion for $X(u)$ and therefore $x(t)$.

Consider, then, the generalized integral transform[2]

$$y(t) = \int_I Y(u)C(t; u)du \qquad (4.83)$$

where $C(t; u)$ is a given kernel, and I a given interval. [For the specific case of Fourier series, $C(t; u) = \exp(j2\pi ut) / \sqrt{2B}$

---

[2]Clearly, $y(t)$ and $Y(u)$ here are not Fourier transform pairs.

and $I = \{u| - B < u < B\}]$. Assume that over the interval I, the functions $\{C(t_n; u) \mid -\infty < n < \infty\}$ form a complete orthonormal basis set which can be used to express $Y(u)$. Then $Y(u)$ can be expressed in an orthonormal expansion using samples of $y(t)$ as coefficients:

$$Y(u) = \sum_{n=-\infty}^{\infty} y(t_n) \, C^*(t_n; u).$$

Substituting into (4.83) gives a generalization of the cardinal series

$$y(t) = \sum_{n=-\infty}^{\infty} y(t_n) \, k_n(t) \tag{4.84}$$

where the nth interpolation function is

$$k_n(t) = \int_I C^*(t_n; u) C(t; u) du. \tag{4.85}$$

4.10   Exercises

4.1   Let $x(t)$ have a bandwidth of B. Let $r = B/W \le 1$. Consider the sinc squared interpolation:

$$y(t; A) = D \sum_{n=-\infty}^{\infty} x(nT)[A \operatorname{sinc}\{A(t - nT)\}]^2$$

where D is a constant and $T = 1/2W$. Let C be such that

$$B \le C \le 2W - B.$$

(a)   Find D such that

$$y(t; C) - y(t; B) = x(t).$$

(b)   Find a filter $H(\frac{u}{2B})$ that gives $x(t)$ as an output when $y(t; C)$ is input.

4.2   (a)   Let

$$k(2Bt) = \frac{1}{2B} e^{-at} \mu(t).$$

Restore the resulting generalized interpolation in (4.21) using a differentiator, a low pass filter and an amplifier.

(b)   Same except

$$k(2Bt) = \frac{1}{2B} e^{-a|t|}.$$

Here, you are allowed an inverter, two amplifiers, two differentiators and a low pass filter for restoration.

4.3   (a)   Show that a signal's Hilbert transform can be obtained by passing the signal through a filter with frequency response

$$H(u) = -j \operatorname{sgn}(u).$$

(b) Let $f(t)$ be a bandlimited signal with bandwidth B and let $g(t)$ be its Hilbert transform. Find $f(t)$ from $\{g\left(\frac{n}{2B}\right) \mid -\infty < n < \infty\}$.

4.4 A bandlimited signal, $x(t)$, and its Hilbert transform are both sampled in phase at half their Nyquist rates. Generate the interpolation functions required to regain $f(t)$.

4.5 Generate an alternate method for restoring lost samples by evaluating (4.11) at the points $\{t = \frac{n}{2W} \mid n \notin \mathcal{M}\}$.

4.6 Except for $n = \pm 1$, a signal's samples are

$$f(nT) = \begin{cases} \frac{4(-1)^{n/2}}{\pi(1-n^2)} & ; \quad n \text{ even} \\ 0 & ; \quad \text{otherwise.} \end{cases}$$

Given $r = 1/2$, find $f(\pm T)$.

4.7 Show that the equations in section 4.1.1.1. are valid for any r in the interval $B/W < r < 1$.

4.8 Except for $n = 0$, a signal's samples are

$$f(nT) = \begin{cases} \frac{2}{n}(-1)^{n/2} & ; \quad \text{even } n \\ \frac{4(-1)^{(n+1)/2}}{\pi n^2} & ; \quad \text{odd } n \end{cases}.$$

The signal is known to be oversampled, but the value of r is uncertain. Find $f(0)$.

4.9 The determinant of the matrix

$$\begin{bmatrix} 1 & x_1 & x_1^2 & \cdots & x_1^{N-1} \\ 1 & x_2 & x_2^2 & \cdots & x_2^{N-1} \\ \vdots & \vdots & \vdots & \cdots & \vdots \\ 1 & x_N & x_N^2 & \cdots & x_N^{N-1} \end{bmatrix}$$

is called the *Vandermonde determinant* and is equal to

$$\Delta = \Pi_{1 \leq j < k \leq N}(x_k - x_j).$$

For example, for $N = 4$,

$$\Delta = (x_4 - x_3)(x_4 - x_2)(x_4 - x_1)$$
$$\times (x_3 - x_2)(x_2 - x_1)$$
$$\times (x_2 - x_1).$$

Use this result to derive (4.47) by using (4.46) in (4.45) with Cramer's rule.

4.10 Let $f(t)$ have bandwidth B. The signals $f(t - \alpha)$ and $f(t + \alpha)$ are sampled uniformly at a rate of B. Show that [Papoulis (1977)]:

$$f(t) = \frac{\cos(2\pi B\alpha) - \cos(2\pi Bt)}{2\pi B \; \sin(2\pi B\alpha)}$$
$$\times \sum_{n=-\infty}^{\infty} \frac{f(\frac{n}{B} + \alpha)}{B(t - \alpha) - n} - \frac{f(\frac{n}{B} - \alpha)}{B(t + \alpha) - n}.$$

4.11 (a) Derive the interpolation functions in (4.53) and (4.54).

(b) Show that the formula in (4.55) not only interpolates the signal samples properly, but also interpolates the derivative samples.

4.12 Why can't we allow $M = N$ in (4.74) ?

4.13 Show that the Fourier coefficients of a trigonometric polynomial can be generated directly from the signal's samples by the matrix equation

$$\vec{c} = \mathbf{A} \, \vec{x}$$

where $\vec{x}$ contains the P signal samples, $\vec{c}$ contains the $2N + 1$ Fourier coefficients and the nq^{th} element of $\mathbf{A}$ is

$$a_{nq} = \int_{-\frac{1}{2}}^{\frac{1}{2}} \frac{\sin[\pi(Pt - q)]/P}{\sin[\pi(Pt - q)/P]} \, e^{j2\pi nt} \, dt.$$

4.14 Let $v(t)$ denote a real baseband signal with a maximum frequency component of $B/2$. The signal

$$x(t) = v(t)\cos(2\pi f_0 t)$$

is bandpass. In section 4.6.1, we showed that $x(t)$, when heterodyned to baseband, required a minimum sampling rate of $2B$. Show a technique whereby a down heterodyned version of our $x(t)$ requires a sampling rate of half that much $(f_0 > B)$.

4.15 *Implicit sampling* of a function $x(t)$ is illustrated in Fig.4.18. A sample is taken when $x(t)$ crosses a predetermined level. Assume that the levels are each separated by an interval of Δ and that one of the levels is at zero. Show that not all finite energy bandlimited signals are determined uniquely by their implicit samples for any finite value of Δ.
HINT: Assume an average sampling density of $2B$ is necessary to uniquely specify the signal and consider the function

$$y(t) = \text{sinc}^2(t) + \text{sinc}^2(t - a)$$

which is strictly positive when a is not an integer.

4.16 Does (4.79) satisfy the criterion for an interpolation function? If not, why?

4.17 What class of functions does Lagrangian interpolation always interpolate exactly using only N samples?

4.18 Derive a closed form expression for the interpolation function for recurrent nonuniform sampling using Lagrangian interpolation. Is it the same as (4.47)?

4.19 Does Lagrangian interpolation result in the expression in (4.2) for oversampled signals?

4.20 Show that for any oversampled bandlimited function that

$$x(t) = r \sum_{n \neq 0} x(\frac{n}{2W}) \; [\text{sinc}(2Bt - rn) +$$

$$\frac{r}{(1-r)} \text{sinc}(rn) \, \text{sinc}(2Bt)]. \quad (4.86)$$

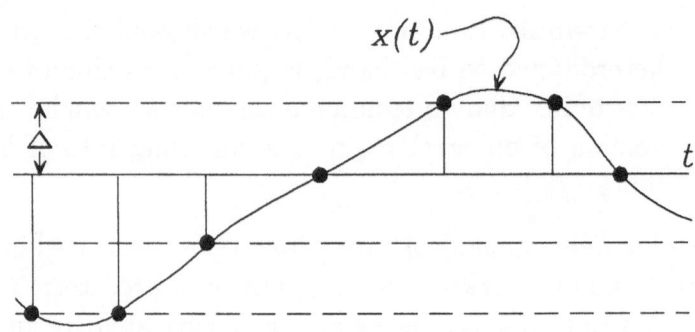

Figure 4.18

4.21 A bandpass function with bandwidth B is directly sampled at a rate $2B$ where B is an integer multiple of f_L. The samples are interpolated using the conventional cardinal series. Outline the processing required to regain the original signal. Does it make a difference whether the integer multiple is odd or even?

4.22 For the filters in (4.56), derive the corresponding interpolation functions for $N = 3$.

4.23 For a given B, let $I = \{u| - B < u < B\}$. Let

$$C(t; u) = \{\text{sgn}[\cos(2\pi ut)] + j\, \text{sgn}[\sin(2\pi ut)]\} / \sqrt{2B}.$$

The set of functions, $\{C(\frac{n}{2B}; u)| - \infty < n < \infty\}$, (known as *complex Walsh functions*), form a complete orthonormal basis set for finite energy functions on the interval I. Evaluate the interpolation functions, $k_n(t)$, corresponding to Kramer's generalization of the sampling theorem for this basis set.

REFERENCES

M. Abramowitz and I.A. Stegun, **Handbook of Mathematical Functions**, Dover, New York, 9th.ed., 1964.

N.C. Gallagher Jr. and G.L. Wise "A representation for bandlimited functions", *Proc. IEEE*, vol. 63, p.1624 (1975).

I.S. Gradshteyn and I.M. Ryzhik, **Tables of Integrals, Products and Series**, Academic Press, New York, 4th ed.,1965.

H.P. Kramer "A generalized sampling theorem", *J. Math. Phys.*, vol. 38, pp.68-72 (1959).

D.A. Linden "A discussion of sampling theorems", *Proc. IRE*, vol. 47, pp.1219-1226 (1959).

D.A.Linden and N.M. Abramson "A generalization of the sampling theorem", *Inform. Contr.*, vol.3, pp.26-31 (1960).

D.G. Luenberger, **Optimization by Vector Space Methods**, Wiley, New York, 1969.

R.J. Marks II and M.W. Hall "Differintegral interpolation from a bandlimited signal's samples", *IEEE Transactions on Acoustics, Speech and Signal Processing*, vol. ASSP-29, pp.872-877 (1981).

R.J. Marks II "Restoring lost samples from an oversampled bandlimited signal", *IEEE Transactions on Acoustics, Speech and Signal Processing*, vol. ASSP-31, pp.752-755 (1983).

A.W. Naylor and G.R. Sell, **Linear Operator Theory in Engineering and Science**, Springer-Verlag, New York, 1982.

A. Papoulis, **Signal Analysis**, McGraw-Hill, New York, 1977.

A. Papoulis "Generalized sampling expansion", *IEEE Transactions on Circuits and Systems*, vol. CAS-24, pp.652-654 (1977).

A. Ralston and P. Rabinowitz, **A First Course in Numerical Analysis, 2nd Ed.**, McGraw-Hill, New York, 1978.

C.E. Shannon "A mathematical theory of communication", *Bell System Technical Journal*, vol. 27, pp.379, 623, (1948).

C.E. Shannon "Communications in the presence of noise", *Proc. IRE*, vol. 37, pp.10-21 (1949).

J.L. Yen "On the nonuniform sampling of bandwidth limited signals", *IRE Transactions on Circuit Theory*, vol. CT-3, pp.251-257 (1956).

5

Sources of Error

Exact interpolation using the cardinal series assumes that (a) the values of the samples are known exactly, (b) the sample locations are known exactly and (c) an infinite number of terms are used in the series. Deviation from these requirements results in interpolation error due to (a) data noise (b) jitter and (c) truncation respectively. The perturbation to the interpolation from these sources of error is the subject of this chapter.

5.1 Effects of Additive Data Noise

If noise is superimposed on sample data, the corresponding interpolation will be perturbed. In this section, the nature of this perturbation is examined. The effect of data noise on continuous sampling interpolation is treated in Chapter 7.

5.1.1 On Cardinal Series Interpolation

Suppose that the signal we sample is corrupted by real additive zero mean wide sense stationary noise, $\xi(t)$. Then, instead of sampling the deterministic bandlimited signal, $x(t)$, we would be sampling the signal

$$y(t) = x(t) + \xi(t). \tag{5.1}$$

From these samples, we form the series

$$z(t) = \sum_{n=-\infty}^{\infty} y(\tfrac{n}{2W})\text{sinc}(2Wt - n) \tag{5.2}$$

where the sampling rate, $2W$, equals or exceeds twice the bandwidth, B, of $x(t)$. Recall the sampling rate parameter

$$r = \frac{B}{W} \le 1.$$

In general, $z(t)$ will equal $y(t)$ only at the sample point locations. Substituting (5.1) into (5.2) reveals that

$$z(t) = x(t) + \eta(t) \tag{5.3}$$

where

$$\eta(t) = \sum_{n=-\infty}^{\infty} \xi(\tfrac{n}{2W}) \operatorname{sinc}(2Wt - n). \tag{5.4}$$

Therefore, $\eta(t)$ is the stochastic process generated by the samples of $\xi(t)$ alone and is independent of the signal. Note that, since $\xi(t)$ is zero mean, so is $\eta(t)$. Hence, expectating both sides of (5.3) leads us to the desirable conclusion that

$$\overline{z(t)} = x(t).$$

5.1.1.1   Interpolation Noise Level

A meaningful measure of the cardinal series' noise sensitivity is the *interpolation noise level* which, since $\xi(t)$ is zero mean, is the noise variance, $\overline{\eta^2(t)}$. Towards this end, we will first find the autocorrelation for $\eta(t)$. From (5.4)

$$
\begin{aligned}
R_\eta(t - \tau) &= E\left[\, \eta(t)\eta(\tau) \,\right] \\
&= \sum_{n=-\infty}^{\infty} \sum_{m=-\infty}^{\infty} R_\xi\left(\frac{n-m}{2W}\right)\operatorname{sinc}(2Wt - n)\operatorname{sinc}(2W\tau - m) \\
&= \sum_{k=-\infty}^{\infty} R_\xi(\tfrac{k}{2W})\sum_{n} \operatorname{sinc}(2W\tau - n+k)\operatorname{sinc}(2Wt - n) \quad (5.5)
\end{aligned}
$$

where our assumption of the wide sense stationarity of $\eta(t)$ will shortly be justified and, in the second step, we have let $k = n - m$. The n sum in (5.5) can be evaluated using the cardinal series applied to $x(t) = \operatorname{sinc}[2W(\tau - t) + k]$. Then

$$R_\eta(t) = \sum_{k=-\infty}^{\infty} R_\xi(\tfrac{k}{2W})\operatorname{sinc}(2Wt - k). \tag{5.6}$$

Thus, the interpolation noise autocorrelation is found from the cardinal series interpolation using sample values from the data noise autocorrelation.

To find the interpolation noise level, we simply evaluate (5.6) at $t = 0$. The remarkable result is:

$$\overline{\eta^2} = \overline{\xi^2}. \tag{5.7}$$

That is, the cardinal series interpolation results in a noise level equal to that of the original signal before sampling [Bracewell].

5.1.1.2   Effects of Oversampling and Filtering

In many cases, one can reduce the interpolation noise level by oversampling and filtering. If, for example, we place $z(t)$ in (5.3) through a filter that is unity for $\mid u \mid < B$ and zero otherwise, then the output is

$$
\begin{aligned}
z_r(t) &= z(t) * 2B\mathrm{sinc}(2Bt) \\
&= x(t) + \eta_r(t)
\end{aligned}
$$

where the stochastic process, $\eta_r(t)$, is defined as

$$\eta_r(t) = \eta(t) * 2B\,\mathrm{sinc}(2Bt).$$

We now will show that

$$\overline{\eta_r^2} \leq \overline{\eta^2} = \overline{\xi^2}. \tag{5.8}$$

That is, filtering reduces or maintains the interpolation noise level since power due to the high frequency components of the noise is eliminated.

Using the appropriate form of (2.38) with $h(t) = 2B\mathrm{sinc}(2Bt)$ gives

$$S_{\eta_r}(u) = S_\eta(u)\,\Pi(\tfrac{u}{2B}).$$

From (2.29),

$$\overline{\eta_r^2} = \int_{-B}^{B} S_\eta(u)\,du \tag{5.9}$$

whereas

$$\overline{\eta^2} = \int_{-W}^{W} S_\eta(u)du. \qquad (5.10)$$

Since power spectral densities are non negative, comparison of (5.9) and (5.10) immediately reveals that the noise level is maintained or reduced as was advertised in (5.8).

We now investigate this reduction more specifically for two types of noise autocorrelations.

(a) Discrete White Noise

Here, we assume that

$$R_\xi(\tfrac{n}{2W}) = \overline{\xi^2}\delta[n].$$

Then, from (5.6):

$$R_\eta(t) = \overline{\xi^2}\text{sinc}(2Wt).$$

Thus

$$S_\eta(u) = \tfrac{\overline{\xi^2}}{2W} \, \Pi(\tfrac{u}{2W})$$

and, from (5.9)

$$\overline{\eta_r^2} = \tfrac{\overline{\xi^2}}{2W} \int_{-B}^{B} \Pi(\tfrac{u}{2W})du$$

$$= r\overline{\xi^2}. \qquad (5.11)$$

The noise level is reduced by the ratio of the Nyquist to the sampling rate.

(b) Laplace Autocorrelation

If the data noise has a Laplace autocorrelation with parameter λ as in (2.33), then the Fourier transform of (5.6) is

$$S_\eta(u) = \tfrac{\overline{\xi^2}}{2W} \sum_{n=-\infty}^{\infty} e^{\frac{-\lambda|n|}{2W}} e^{-j\pi nu/W} \, \Pi(\tfrac{u}{2W})$$

$$= \tfrac{\overline{\xi^2}}{2W} \sum_{n=-\infty}^{\infty} e^{\frac{-\lambda|n|}{2W}} \cos(\pi nu/W) \, \Pi(\tfrac{u}{2W})$$

where, in the second step, we have recalled that S_η is real. Continuing:

$$S_\eta(u) = \frac{\overline{\xi^2}}{2W}[1 + 2\sum_{n=1}^{\infty} e^{\frac{-\lambda n}{2W}} \cos(\pi n u/W)] \Pi(\tfrac{u}{2W})$$

$$= \frac{\overline{\xi^2}}{2W}[1 + 2\Re\sum_{n=1}^{\infty} e^{\frac{-n(\lambda+j2\pi u)}{2W}}] \Pi(\tfrac{u}{2W}). \qquad (5.12)$$

Recall the geometric series:

$$\sum_{n=0}^{\infty} z^n = (1 - z)^{-1}; |z| < 1. \qquad (5.13)$$

Applying to (5.12) and simplifying gives the (unfiltered) interpolation noise power spectral density [Marks (1983)]:

$$S_\eta(u) = \frac{\overline{\xi^2}}{2W} \frac{\sinh(\tfrac{\lambda}{2W}) \Pi(\tfrac{u}{2W})}{\cosh(\tfrac{\lambda}{2W}) - \cos(\pi u/W)}.$$

The power spectral density for $\eta_r(t)$ is the same, but is only nonzero over the interval $|u| < B$. The filtered interpolation noise level, from (5.9), follows as

$$\overline{\eta_r^2} = \overline{\xi^2} \sinh(\tfrac{\lambda}{2W}) \int_0^r [\cosh(\tfrac{\lambda}{2W}) - \cos(\pi\nu)]^{-1} d\nu \qquad (5.14)$$

where we have made the variable substitution $\nu = 2uT$ and have recognized the integrand is even. Since [Gradshteyn & Ryzhik]

$$\int \frac{d\gamma}{a + b\cos(\gamma)} = \frac{2}{\sqrt{a^2 - b^2}} \arctan[\frac{\sqrt{a^2 - b^2}\tan(\gamma/2)}{a + b}] ; a^2 > b^2$$

equation (5.14) can be evaluated as

$$\overline{\eta_r^2} = \frac{2\overline{\xi^2}}{\pi} \arctan[\frac{\sinh(\tfrac{\lambda}{2W})\tan(\tfrac{\pi r}{2})}{\cosh(\tfrac{\lambda}{2W}) - 1}]. \qquad (5.15)$$

Since the principle value of the arctan is strictly less than $\pi/2$, it is clear that the filtered interpolation noise level is less than the data noise level.

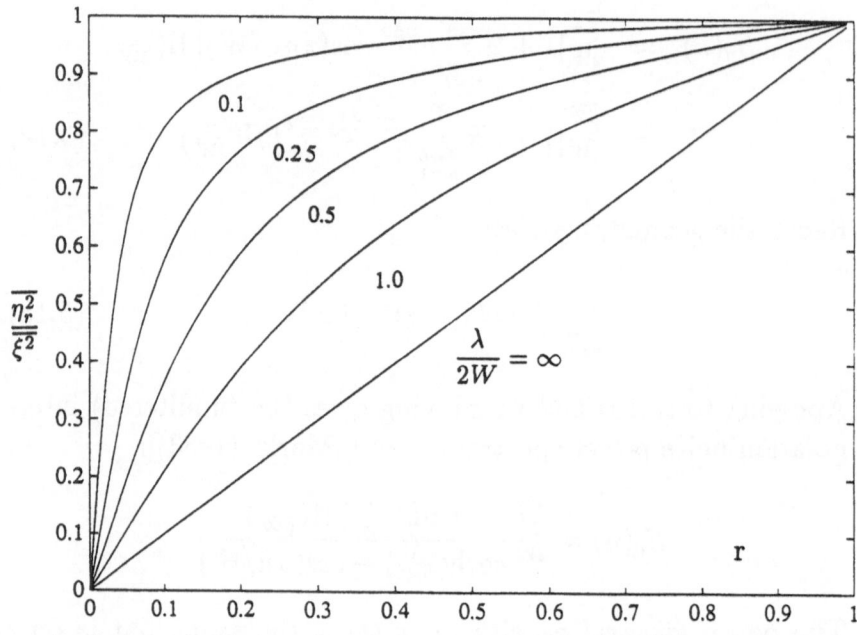

Figure 5.1: Plots of interpolation noise variance for additive data noise with Laplace autocorrelation with parameter λ.

Note that for large λ, the Laplace autocorrelation approaches the autocorrelation of discrete white noise. This follows from

$$\lim_{\rho \to \infty} \frac{\sinh(\rho)}{\cosh(\rho) - 1} = 1.$$

The corresponding limiting case of (5.15) is thus the same as that for white noise in (5.11). Note also for $r = 1$ and η replacing η_r, that (5.15) reduces to (5.7).

Plots of $\overline{\eta_r^2}/\overline{\xi^2}$ are shown in Fig. 5.1 as a function of r for various values of $\lambda/2W$. The higher the correlation between adjacent noise samples, the higher the filtered interpolation noise level.

5.1.2   Interpolation Noise Variance for Directly Sampled Signals

The results of the previous section can be nicely generalized to the sampling theorems listed in Table 4.1, all of which can be written as

$$x(t) = \sum_{n \in S} x(t_n) k_n(t) \qquad (5.16)$$

where S is a set of integers. Suppose the data were corrupted by real additive zero mean stationary noise $\xi(t)$. Then $x(t_n) + \xi(t_n)$ would appear in the summand of (5.16) rather than just $x(t_n)$. The result is clearly $x(t) + \eta(t)$ where the interpolation noise is

$$\eta(t) = \sum_{n \in S} \xi(t_n) k_n(t). \qquad (5.17)$$

Our noisy interpolated signal is then

$$z(t) = x(t) + \eta(t). \qquad (5.18)$$

Since $\xi(t)$ is zero mean, we have the desirable property that

$$\overline{z(t)} = x(t).$$

The second order statistics of $\eta(t)$ reveal the uncertainty of our estimate. Using (5.17), we have

$$\begin{aligned} R_\eta(t; \tau) &= E[\eta(t)\eta(\tau)] \\ &= \sum_{n \in S} \sum_{m \in S} R_\xi(t_n - t_m) k_n(t) k_m(\tau). \end{aligned} \qquad (5.19)$$

The interpolation noise variance follows as

$$\overline{\eta^2(t)} = R_\eta(t; t). \qquad (5.20)$$

If there is oversampling, the signal, once interpolated, can be filtered to remove noise components not in the pass band. If the signal is low pass with bandwidth B, the result is

$$\begin{aligned} z_r(t) &= z(t) * 2B \operatorname{sinc}(2Bt) \\ &= x(t) + \eta_r(t) \end{aligned}$$

where

$$\eta_r(t) = \eta(t) * 2B \operatorname{sinc}(2Bt).$$

Discrete White Noise: The expressions for the second order statistics simplify significantly if the noise samples $\{\xi(t_n) \mid n \in S\}$ are uncorrelated (white). Then

$$R_\xi(t_n - t_m) = \overline{\xi^2}\delta[n - m], \tag{5.21}$$

and (5.19) reduces to a single sum:

$$R_\eta(t; \tau) = \overline{\xi^2} \sum_{n \in S} k_n(t)k_n(\tau), \tag{5.22}$$

and the *normalized interpolation noise variance* (NINV) becomes

$$\overline{\eta^2(t)}/\overline{\xi^2} = \sum_{n \in S} k_n^2(t). \tag{5.23}$$

If the interpolated signal is filtered, the resulting NINV is:

$$\overline{\eta_r^2(t)}/\overline{\xi^2} = \sum_{n \in S} [k_n^{(r)}(t)]^2 \tag{5.24}$$

where, if the signal is low pass with bandwidth B,

$$k_n^{(r)}(t) = k_n(t) * 2B \operatorname{sinc}(2Bt). \tag{5.25}$$

The general results of this section will now be applied to some specific cases.

5.1.2.1   Interpolation with Lost Samples

We here consider the NINV resulting from interpolation in the presence of lost samples [Marks and Radbel]. We will demonstrate that the NINV increases when (a) the $r < 1$ sampling rate becomes close to that of Nyquist, (b) the number of lost samples increases and/or (c) the lost sample locations are "close." Analysis will be restricted to additive white noise as in (5.21).

(a) One Lost Sample

For one lost sample at the origin, we use the corresponding interpolation function in (4.10):

$$k_n(t) = \text{sinc}(2Wt - n) + \frac{r}{1-r}\text{sinc}(rn)\text{sinc}(2Wt). \qquad (5.26)$$

Substituting into (5.23) gives

$$\overline{\eta^2(t)}/\overline{\xi^2} = \sum_{n\neq 0}[\,\text{sinc}(2Wt - n) + \frac{r}{1-r}\text{sinc}(rn)\text{sinc}(2Wt)]^2$$

$$= \sum_{n=-\infty}^{\infty}[\,\text{sinc}^2(2Wt - n) + (\frac{r}{1-r})^2\text{sinc}^2(rn)\text{sinc}^2(2Wt)$$

$$+ \frac{2r}{1-r}\text{sinc}(2Wt - n)\text{sinc}(rn)\text{sinc}(2Wt)\,]$$

$$- \frac{1}{(1-r)^2}\text{sinc}^2(2Wt). \qquad (5.27)$$

Each of the three infinite sums can be evaluated in closed form. For the first term, we set $\tau = t$ in the cardinal series expansion

$$\text{sinc}2W(t - \tau) = \sum_{n=-\infty}^{\infty}\text{sinc}(2W\tau - n)\,\text{sinc}(2Wt - n).$$

The cardinal series applied to

$$\text{sinc}(2Bt) = \sum_{n=-\infty}^{\infty}\text{sinc}(rn)\,\text{sinc}(2Wt - n) \qquad (5.28)$$

lets us evaluate the third sum and the series in (4.2) applied to

$$\text{sinc}(2Bt) = r\sum_{n=-\infty}^{\infty}\text{sinc}(rn)\,\text{sinc}(2Bt - rn) \qquad (5.29)$$

(set $t = 0$) gives the second sum . Alternately, (5.29) is a low passed version of (5.28). Collecting terms and simplifying leaves

$$\overline{\eta^2(t)}/\overline{\xi^2} = 1 + \frac{2r}{1-r}\text{sinc}(2Wt)\,\text{sinc}(2Bt)$$

$$- \frac{1}{1-r}\text{sinc}^2(2Wt). \qquad (5.30)$$

Note that, for large t, the NINV approaches unity. This is consistent with (5.7) since, far removed from the origin, the effect of the lost sample is negligible.

The noise at the origin follows from (5.30) as

$$\overline{\eta^2(0)}/\overline{\xi^2} = \frac{r}{1-r}. \qquad (5.31)$$

The result is monotonically increasing on $0 < r \leq 1$. Interestingly, for $r < 1/2$, the normalized interpolation noise level in (5.31) is less than unity which is less than the noise level of the known sample data. Note, however, that we have yet to filter the high-frequency components of the discrete white noise.

For the filtered case for one lost sample, the interpolation function, from Table 4.1, is

$$k_n^{(r)}(t) = r \operatorname{sinc}(2Bt - rn) + \frac{r^2}{1-r} \operatorname{sinc}(rn)\operatorname{sinc}(2Bt).$$

From (5.23), the corresponding NINV is:

$$\overline{\eta_r^2}/\overline{\xi^2} = r^2 \sum_{n \neq 0} [\operatorname{sinc}(2Bt - rn) + \frac{r}{1-r}\operatorname{sinc}(rn)\operatorname{sinc}(2Bt)]^2.$$

Proceeding in a manner similar to that for the unfiltered case above, we obtain

$$\overline{\eta_r^2}/\overline{\xi^2} = \frac{r}{1-r}[1 - r\{1 - \operatorname{sinc}^2(2Bt)\}]. \qquad (5.32)$$

For large t, the noise level goes to the no lost sample filtered equivalent in (5.11). Note, in particular, from (5.31) that

$$\overline{\eta_r^2(0)} = \overline{\eta^2(0)}.$$

Hence, filtering the interpolation does not improve the uncertainty of the restoration of the lost sample. As we would expect from (5.7) and (5.11) respectively,

$$\overline{\eta^2(\pm\infty)} = \overline{\xi^2}$$

and

$$\overline{\eta_r^2(\pm\infty)} = r\overline{\xi^2}.$$

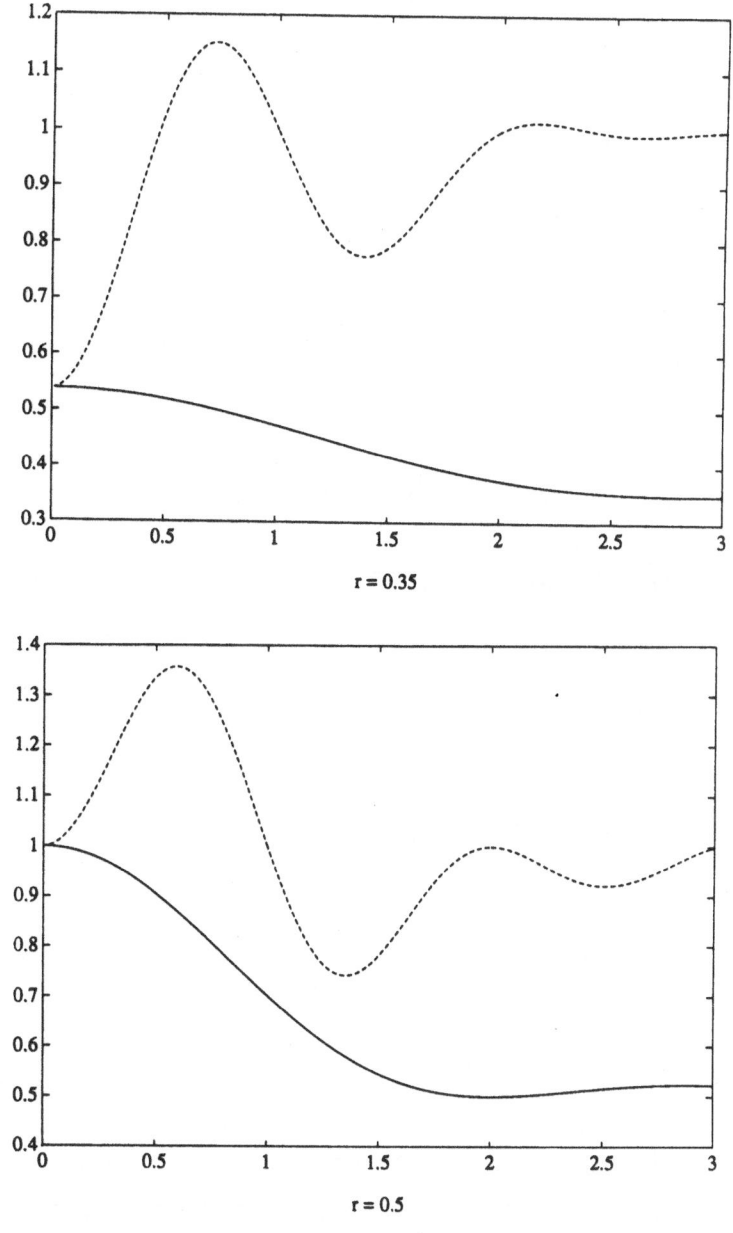

Figure 5.2: Plots of the normalized interpolation noise variance (NINV) for the case of a single lost sample vs. $t/2W$ for four different values of r. The dashed line is for the unfiltered case and the solid is the noise level after filtering.

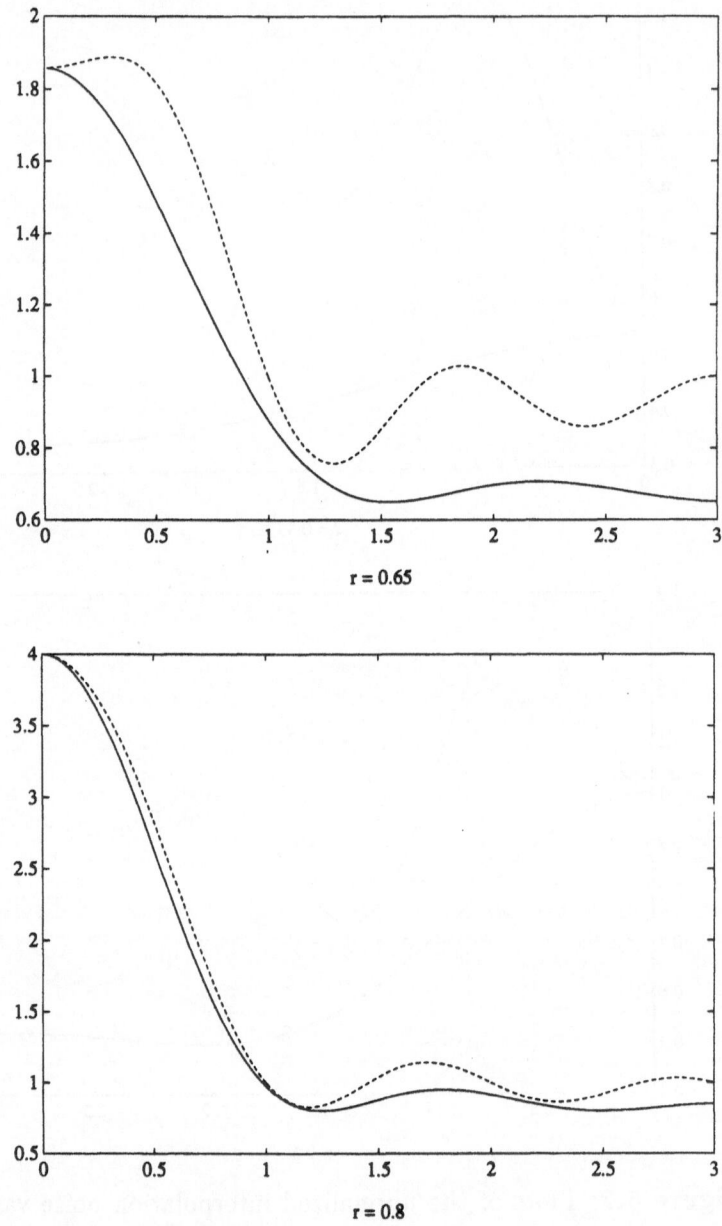

r = 0.65

r = 0.8

Figure 5.2: Continuation of Fig. 5.2.

Plots of (5.30) and (5.32) are shown in Fig. 5.2.

(b) Two Lost Samples

Let $M = 2$ and let the lost samples be located at the origin and at $x = k/2B$ for some specified positive integer k. The 2×2 matrix, $\mathbf{A}$, discussed in Section 4.1.1.1.3, has elements

$$a_{11} = a_{22} = \frac{1-r}{\Delta}$$

and

$$a_{12} = a_{21} = \frac{r \, \text{sinc}(rk)}{\Delta}$$

where

$$\Delta = (1-r)^2 - r^2 \, \text{sinc}^2(rk).$$

The corresponding interpolation functions in (4.16) and (4.18) are substituted into (5.23) and (5.24), respectively. After straightforward yet tedious calculations, we obtain

$$
\begin{aligned}
\overline{\eta^2(t)}/\overline{\xi^2} \;=\; & 1 - (\alpha^2 + \beta^2) + 2r\,[\,a_{11}(\alpha\tau + \beta\rho) + a_{12}(\alpha\rho + \beta\tau)\,] \\
& + r^2\,[\,(a_{11}^2 + a_{12}^2)(\alpha^2\lambda + 2\alpha\beta\gamma + \beta^2\lambda) \\
& + 2a_{11}a_{12}(\alpha^2\gamma + 2\alpha\beta\lambda + \beta^2\gamma)\,]
\end{aligned}
\tag{5.33}
$$

and

$$
\begin{aligned}
\overline{\eta_r^2(t)}/\overline{\xi^2} \;=\; & r - r^2(\alpha_1^2 + \beta_1^2) \\
& + 2r^3\,[\,a_{11}(\alpha_1\tau_1 + \beta_1\rho_1) + a_{12}(\alpha_1\rho_1 + \beta_1\tau_1)\,] \\
& + r^4[\,(a_{11}^2 + a_{12}^2)(\alpha_1^2\lambda + 2\alpha_1\beta_1\gamma + \beta_1^2\lambda) \\
& + 2a_{11}a_{12}(\alpha_1\gamma + 2\alpha_1\beta_1\lambda + \beta_1\gamma)\,]
\end{aligned}
\tag{5.34}
$$

where

$$
\begin{aligned}
\alpha &= \text{sinc}(2Wt - k), & \alpha_1 &= \text{sinc}(2Bt - rk) \\
\beta &= \text{sinc}(2Wt), & \beta_1 &= \text{sinc}(2Bt)
\end{aligned}
$$

and

$$\rho = \sum_{p \neq 0,k} \text{sinc}(rp)\text{sinc}(2Wt - p)$$

$$= \beta_1 - \beta - \alpha \operatorname{sinc}(rk),$$

$$\rho_1 = \sum_{p \neq 0, k} \operatorname{sinc}(rp)\operatorname{sinc}(2Bt - rp)$$

$$= (\frac{1-r}{r})\beta_1 - \alpha_1 \operatorname{sinc}(rk),$$

$$\tau = \sum_{p \neq 0, k} \operatorname{sinc}[r(p - k)]\operatorname{sinc}(2Wt - p)$$

$$= \alpha_1 - \alpha - \beta \operatorname{sinc}(rk),$$

$$\tau_1 = \sum_{p \neq 0, k} \operatorname{sinc}[r(p - k)]\operatorname{sinc}(2Bt - rp)$$

$$= (\frac{1-r}{r})\alpha_1 - \beta_1 \operatorname{sinc}(rk),$$

$$\lambda = \sum_{p \neq 0, k} \operatorname{sinc}^2(rp)$$

$$= \sum_{p \neq 0, k} \operatorname{sinc}^2[r(p - k)]$$

$$= \frac{1}{r} - 1 - \operatorname{sinc}^2(rk),$$

$$\gamma = \sum_{p \neq 0, k} \operatorname{sinc}(rp)\operatorname{sinc}[r(p - k)]$$

$$= (\frac{1}{r} - 2) \operatorname{sinc}(rk).$$

Numerical examples of (5.33) and (5.34) are shown in Fig. 5.3 for $k = 1$ and 5 with $r = 0.2$. The lost sample locations here are at the minima of the unfiltered noise level curves. A second example for $r = 0.8$ and $k = 1$ is shown in Fig. 5.4. The lost samples are at zero and unity. The filtered and unfiltered curves are indistinguishable near those points.

At the lost sample point locations

$$\overline{\eta^2(0)} = \overline{\eta^2(\frac{k}{2B})} = \overline{\eta_r^2(0)} = \overline{\eta_r^2(\frac{k}{2B})}$$

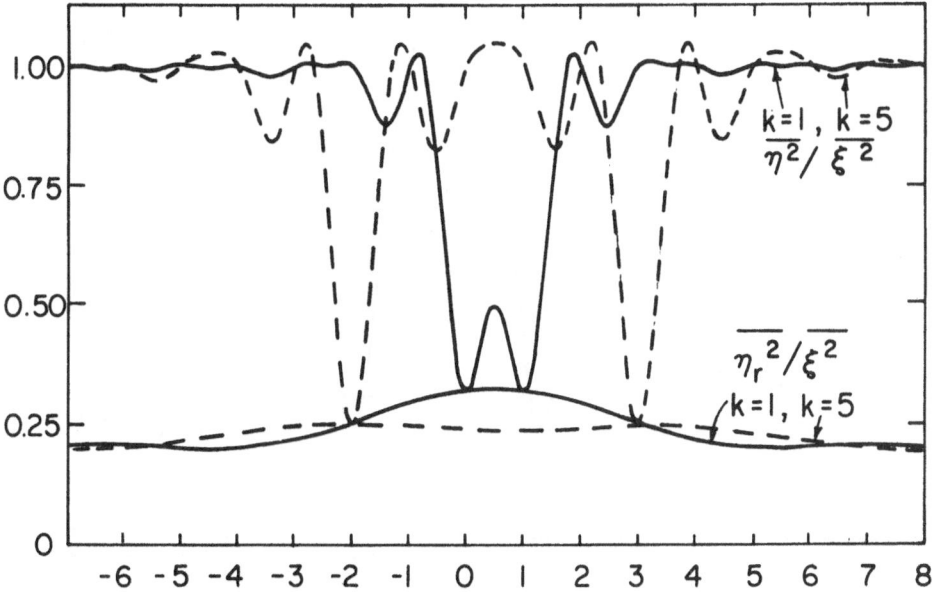

Figure 5.3: NINV for two lost samples as a function of $2Wt$ when
sampling at five times the Nyquist rate ($r = 0.2$).
The solid curves are for $k = 1$. The lost samples
are at zero and one. The broken line graphs are for
$k = 5$ with lost samples at -2 and 3. In both cases,
the lower curve represents the filtered case and the
upper curve the unfiltered case.

$$= \; r^2\overline{\xi^2}\,[(a_{11}^2 + a_{12}^2)\lambda + 2a_{11}a_{12}\gamma]. \qquad (5.35)$$

For large k, the noise level at the origin approaches that for a
single lost sample. If kr is an integer, the noise levels for one
and two lost samples are equal at the lost sample locations. A
plot of (5.35) is shown in Fig. 5.5 for $k = 1, 2$, and 5. The single
lost sample noise level in (5.31) is nearly graphically indistinguishable from the $k = 5$ curve.

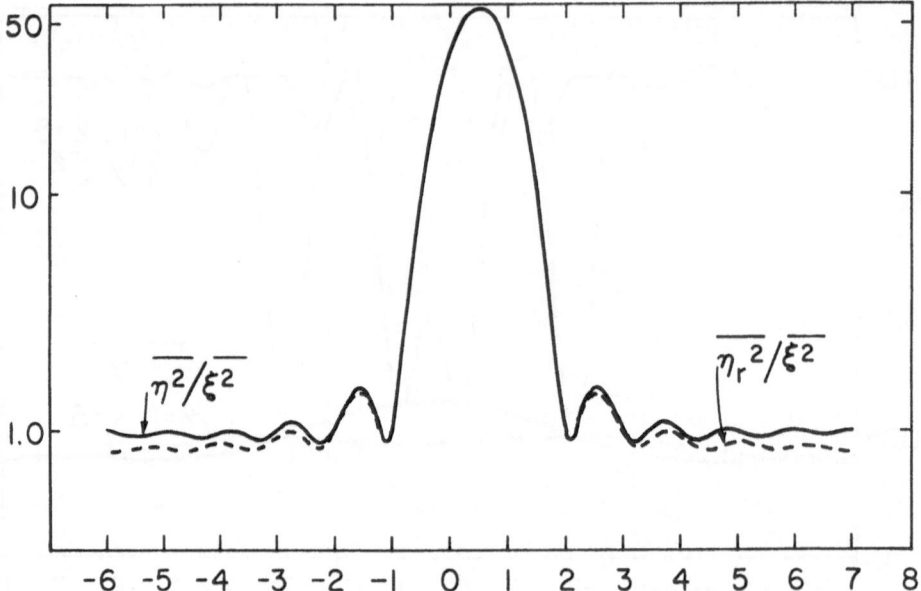

Figure 5.4: NINV for two lost samples as a function of $2Wt$ when $r = 0.8$ and $k = 1$. The lost samples are at zero and one. The solid curve is for the unfiltered case and broken line plot for the filtered case. The two plots are graphically indistinguishable in the region of the lost samples.

(c) A Sequence of Lost Samples

For an even greater number of lost samples, the obtaining of a closed form solution for the NINV using the previous methods becomes nearly intractable. Evaluation of the infinite series numerically becomes more attractive. Alternately, a concise matrix approach to the problem developed by Tseng can be used. We will not, however, review it here.

Numerically evaluated plots of the filtered NINV for three samples in a row are shown in Fig. 5.6 for $r = 0.5$ and 0.8. The NINV at the lost sample locations is shown in Fig. 5.7 for M lost samples in a row. The noise level increases drastically with respect to the number of adjacent lost samples and sampling

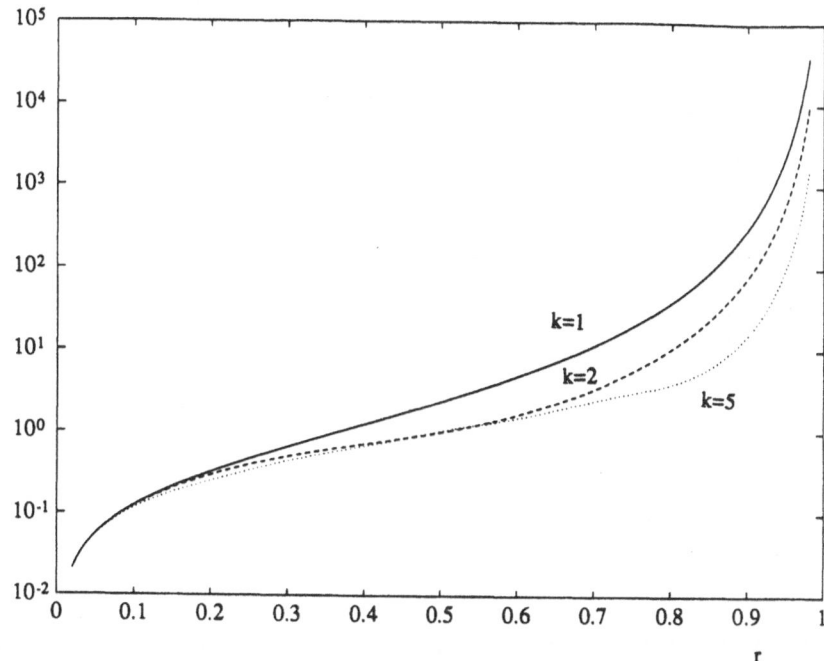

Figure 5.5: NINV for two lost samples at the lost sample location as a function of the sampling rate parameter. The noise level for a single lost sample is almost graphically indistinguishable from the $k = 5$ plot.

rate parameter. Correspondingly, the condition number of the $\mathbf{A} = [\mathbf{I} - \mathbf{S}]^{-1}$ matrix increases greatly with larger M and r.

5.1.2.2   Bandpass Functions

The NINV for bandpass function interpolation follows from Table 4.1 and (5.23) as

$$\overline{\eta^2(t)}/\overline{\xi^2} = \sum_{n=-\infty}^{\infty} \text{sinc}^2(Bt - \frac{n}{2})\cos^2[\pi(2N+1)(Bt - \frac{n}{2})]. \quad (5.36)$$

Expanding the bandpass function

$$x(t) = \text{sinc}B(t - \tau)\cos[\pi(2N + 1)B(t - \tau)]$$

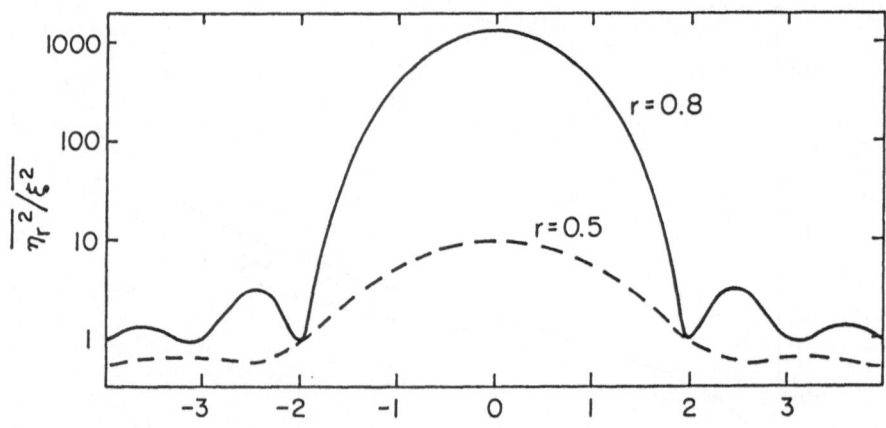

Figure 5.6: Filtered NINV for three lost samples at 0, +1, and
−1 as a function of $2Wt$.

in the bandpass sampling theorem in (4.78) and (4.79) gives

$$\text{sinc}B(t - \tau)\cos[\pi(2N + 1)B(t - \tau)]$$
$$= \sum_{n=-\infty}^{\infty} \text{sinc}(B\tau - \frac{n}{2}) \cos[\pi(2N + 1)(B\tau - \frac{n}{2})]$$
$$\times \text{sinc}(Bt - \frac{n}{2})\cos[\pi(2N + 1)(Bt - \frac{n}{2})].$$

Evaluating this expression at $\tau = t$ reduces (5.36) to

$$\overline{\eta^2(t)}/\overline{\xi^2} = 1.$$

Therefore, as in the cardinal series, the NINV is the same as the
variance of the data noise.

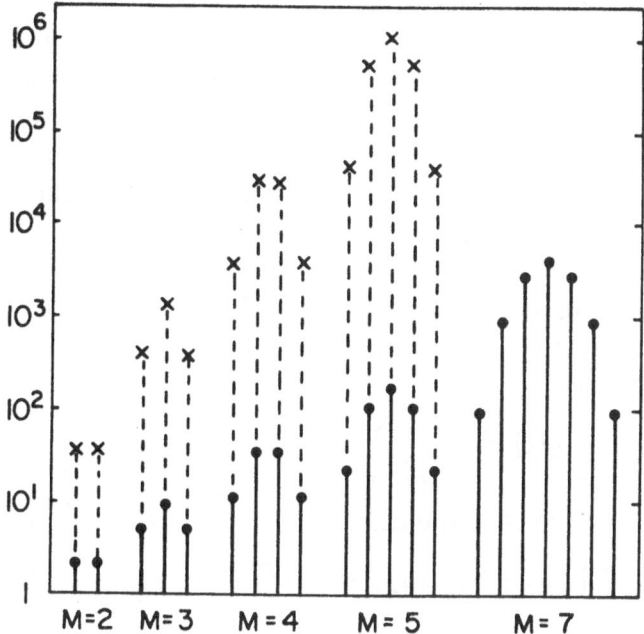

Figure 5.7: NINV of M lost samples in a row at the lost sample locations. The lower dot values in each case correspond to $r = 0.5$ and the upper $\times$'s to $r = 0.8$.

5.1.3   On Papoulis' Generalization

Papoulis' Generalization of the sampling theorem was presented in Sec. 4.2. Here, we will explore the effects of additive white noise superimposed on the sample data. We will expose the *ill-posedness* of a number of innocent appearing sampling theorems [Cheung and Marks]. Interpolation is here defined to be ill-posed if the NINV cannot be bounded. Clearly, such sampling theorems should be avoided.

An example of an ill-posed sampling theorem is a special case of signal-derivative sampling. Shannon was the first to note that one could sample at half the Nyquist rate if at each sample location two samples were taken: one of the signal and one of the signal's derivative. Consider the seemingly innocent alteration of sampling at the Nyquist rate with interlaced signal and first derivative samples taken at each Nyquist interval. As we will demonstrate, restoration here is ill-posed. Indeed, subjecting the samples to sample-wise white noise renders the restoration unstable. Hence, one would wish to sample an odometer and speedometer simultaneously, rather than sequentially, to determine position.

Let $\{\xi_p(nT_N) \mid p = 1, 2, \cdots, N; -\infty < n < \infty\}$ denote a zero mean discrete stochastic noise sequence. If $g_p(nT_N) + \xi_p(nT_N)$ is used in (4.42) instead of $g_p(nT_N)$, the output is $f(t) + \eta(t)$ where

$$\eta(t) = \sum_{p=1}^{N} \sum_{n=-\infty}^{\infty} \xi_p(nT_N) k_p(t - nT_N). \qquad (5.37)$$

We will assume that the discrete noise is stationary and white:

$$E\left[\, \xi_k(nT_N)\xi_p(mT_N)\,\right] = \overline{\xi_p^2}\, \delta[k - p]\, \delta[n - m]$$

where $\overline{\xi_p^2} = E[|\,\xi_p(nT_N)\,|^2]$ is the data noise variance of the pth sampled signal. The interpolation noise level then follows as

$$\overline{\eta^2(t)} = \sum_{p=1}^{N} \overline{\xi_p^2} \sum_{n=-\infty}^{\infty} |\,k_p(t - nT_N)\,|^2 \,.$$

Clearly, $\overline{\eta^2(t)}$ is periodic with period T_N. Application of the Poisson sum formula yields

$$\overline{\eta^2(t)} = \frac{1}{T_N} \sum_{p=1}^{N} \overline{\xi_p^2} \sum_{n=-\infty}^{\infty} W_p(2nB_N)\, e^{j4\pi nB_N t} \qquad (5.38)$$

where

$$|\,k_p(t)\,|^2 \leftrightarrow W_p(u) = \int_{-B}^{B} K_p(\beta)K_p^*(\beta - u)d\beta.$$

Note that (5.38) is simply a Fourier series with coefficients

$$C_n = \frac{1}{T_N} \sum_{p=1}^{N} \overline{\xi_p^2}\, W_p(2nB_N). \qquad (5.39)$$

We accordingly define the average interpolation noise variance by

$$C_0 = \frac{1}{T_N} \sum_{p=1}^{N} \overline{\xi_p^2} \, W_p(0)$$

$$= \frac{1}{T_N} \sum_{p=1}^{N} \overline{\xi_p^2} \int_{-B}^{B} |K_p(u)|^2 \, du \qquad (5.40)$$

or, using Parseval's theorem

$$C_0 = \frac{1}{T_N} \sum_{p=1}^{N} \overline{\xi_p^2} \int_{-\infty}^{\infty} |k_p(t)|^2 \, dt. \qquad (5.41)$$

Thus the average interpolation noise variance is infinite if any one of the N interpolation functions has unbounded energy. Equivalently, if, for any $p = 1, 2, \ldots, N$,

$$\int_{-\infty}^{\infty} |k_p(t)|^2 \, dt = \int_{-B}^{B} |K_p(u)|^2 \, du$$

$$= \infty \qquad (5.42)$$

then the restoration is ill-posed.

5.1.3.1   Examples

1. Derivative Sampling
Consider the $N = 1$ case corresponding to Mth-order derivative sampling

$$H_1(u) = (j2\pi u)^M.$$

We can, in principle, regain all frequency components other than zero. Note, however, that (5.41) becomes

$$C_0 = \frac{\overline{\xi_1^2}}{(2\pi)^{2M} T_N} \int_{-B}^{B} u^{-2M} \, du$$

$$= \infty.$$

The corresponding sampling theorem is thus ill-posed.

2. Interlaced Signal-Derivative Sampling

A less obvious ill-posed sampling theorem occurs when we nonuniformly interlace Mth order derivative samples with signal samples. The sampling theorem for this problem was addressed in Sec. 4.2.3.2. The spectra of the interpolation functions $K_1(u)$ and $K_2(u)$ have real poles when $\Delta(u) = 0$ or $\Delta(u + B) = 0$ on the intervals $(0, B)$ and $(-B, 0)$ respectively. The former occurs when

$$u^M e^{-j2\pi(\alpha B + n)} = (u - B)^M; 0 \le n < M$$

or

$$u = \frac{B}{2}[1 - j\cot\{\pi(\alpha B + n)/M\}].$$

One of these roots is real when (a) $\alpha = 0$ and M is even, or (b) $\alpha = \frac{1}{2B}$ and M is odd (corresponding to $n = M/2$ and $n = (M-1)/2$, respectively). In either case, the real pole generated by $\Delta(u)$ is at $B/2$ and that generated by $\Delta(u + B)$ is at $-B/2$. Clearly, application of (5.40) exposes these sampling theorems as ill-posed. Plots of the spectra of the interpolation functions are shown in Fig. 5.8 for $M = 1$ for various values of α. Fig. 5.9 illustrates the same process for $M = 2$.

5.1.3.2   Notes

1. Sample Contributions in the Ill-Posed Sampling Theorems

Insight into the ill-posedness of the sampling theorems can be gained by inspection of the interpolation functions. Consider, for example, $N = 1$ derivative sampling with $M = 1$. It follows that

$$
\begin{aligned}
k_1(t) &= \frac{1}{4\pi B}\int_{-B}^{B}\frac{e^{j2\pi ut}}{ju}du \\
&= \frac{1}{2\pi B}\text{Si}(2\pi Bt)
\end{aligned}
$$

where $\text{Si}(\cdot)$ is the sine integral. Since $\text{Si}(\pm\infty) = \pm\pi/2$, interpolation at any point is affected significantly by every sample value, no matter how distant.

A similar contribution occurs for the ill-posed cases of interlaced signal-derivative sampling. We can invert (4.51) and

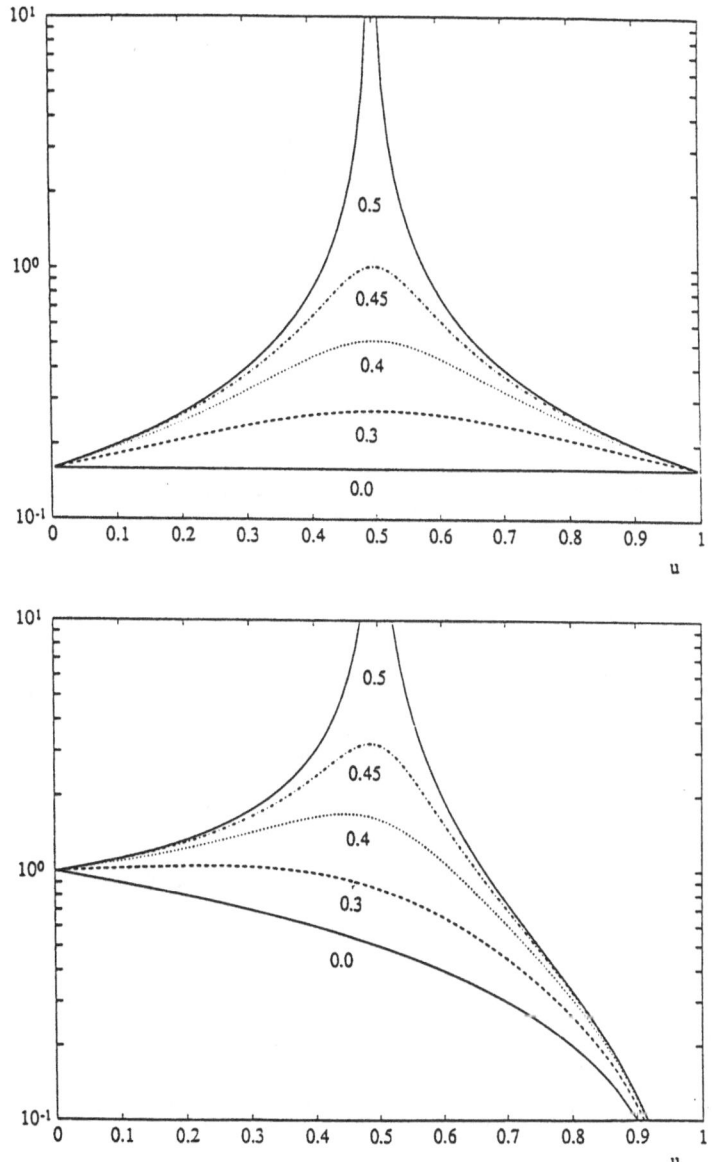

Figure 5.8: Illustration of the manner in which the interpolation functions in (4.51) and (4.52) approach poles for $M = 1$ as $\alpha \to 1/2B$. On top is a plot of $| K_1(u) |$ for different values of α. $| K_2(u) |$ is shown on the bottom. The bandwidth, B, is normalized to unity.

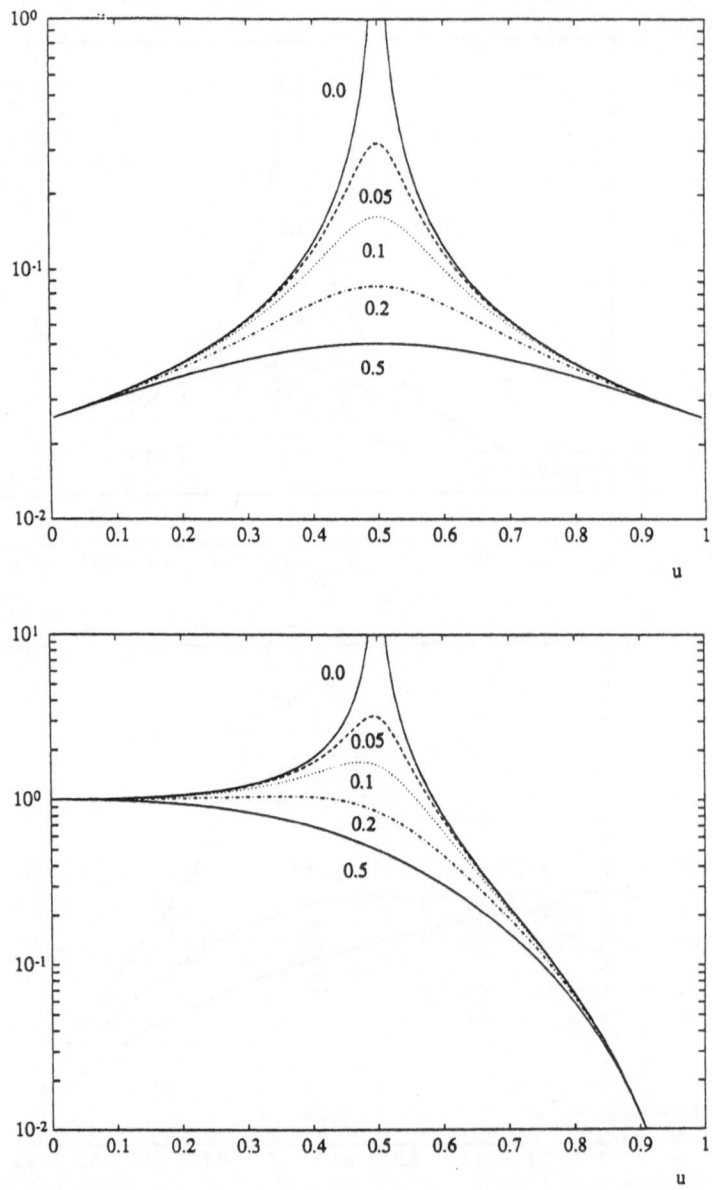

Figure 5.9: Same as Fig. 5.8, except $M = 2$.

(4.52). For $M = 2$ and $\alpha = 0$, the results are

$$k_1(t) = \frac{1}{2}[\sin(\pi Bt)\text{Si}(\pi Bt) + \cos(\pi Bt)\,\text{sinc}(Bt) + \text{sinc}^2(Bt)]$$

and

$$k_2(t) = \frac{1}{2}\frac{\sin(\pi Bt)\text{Si}(\pi Bt)}{(\pi B)^2}.$$

Again, the occurrence of the sine integrals makes possible equally significant contributions from all sample values, no matter how far removed from the point of interpolation. The weighted noise levels from each sample value thus add to a random variable with unbounded variance.

2. Effects of Oversampling

If we sample at a rate $2W > 2B$ and do not take advantage of the oversampling, the average interpolation noise level in (5.40) becomes

$$C_0 = \frac{\overline{\xi^2}}{T_N}\sum_{p=1}^{N}\int_{-W}^{W} | K_p(u) |^2 \, du$$

where, now, $T_N = N/2W$. If, however, the filtered interpolation formula in (4.60) is used, then one can easily show that the average noise variance reduces to:

$$C_0^{(r)} = \frac{\overline{\xi^2}}{T_N}\sum_{p=1}^{N}\int_{-B}^{B} | K_p(u) |^2 \, du. \tag{5.43}$$

Clearly

$$C_0^{(r)} \leq C_0.$$

Thus, as we have seen before, oversampling can reduce interpolation noise variance by allowing suppression of high frequency noise components.

As an example, consider the ill-posed interlaced signal derivative sampling theorem. If we sample at a rate greater than twice the Nyquist rate, the integral in (5.43) will not include the poles at $u = \pm W/2$ and the resulting sampling theorem becomes well-posed. At exactly twice the Nyquist rate, the integration limits in (5.43) are at the pole locations. Thus $C_0^{(r)} = \infty$. We can, however, discard the derivative samples and use the conventional (well-posed) sampling theorem to restore the signal. Thus we are confronted with the curious task of discarding the derivative samples to improve the interpolation noise level.

5.1.4   On Derivative Interpolation

Hamming has noted that "the estimation of derivatives from computed or tabulated values is dangerous." This is largely due to data uncertainty. We will show especially that, even when the noise is bandlimited, NINV's for high order differentiation interpolation can be significantly high [Marks].

Recall from Section 4.3 that derivatives of bandlimited signals can be computed via

$$x^{(p)}(t) = (2W)^p \sum_{n=-\infty}^{\infty} x(\tfrac{n}{2W}) d_p(2Wt - n) \tag{5.44}$$

where the derivative kernel is

$$d_p(t) = (\frac{d}{dt})^p \operatorname{sinc}(t).$$

Since $x^{(p)}(t)$ is bandlimited, it is unaltered by low-pass filtering. Passing (5.44) through a filter unity on $\mid u \mid < B$ and zero elsewhere gives

$$x^{(p)}(t) = r(2B)^p \sum_{n=-\infty}^{\infty} x(\tfrac{n}{2W}) d_p(2Bt - rn). \tag{5.45}$$

If the noise samples, $\xi(\tfrac{n}{2B})$, are added to the signal samples in (5.45), the result is $x^{(p)}(t) + \eta^{(p)}(t)$ where

$$\eta^{(p)}(t) = r(2B)^p \sum_{n=-\infty}^{\infty} \xi(\tfrac{n}{2W}) d_p(2Bt - rn).$$

Thus

$$R_{\eta_p}(t - \tau) = E[\eta^{(p)}(t)\eta^{(p)}(\tau)]$$

$$= r^2(2B)^{2p} \sum_{m=-\infty}^{\infty} R_\xi(\frac{m}{2W}) \sum_{n=-\infty}^{\infty} d_p[2B\tau - (n-m)r] d_p[2Bt - rn].$$

Since $x(t) = d_p[2B(\tau - t) + mr]$ is bandlimited , we can use (5.45) to evaluate the n sum above. Furthermore, since

$$d_p(t) = (-1)^p d_p(-t)$$

and

$$(\frac{d}{dt})^p d_p(t) = d_{2p}(t)$$

it follows that

$$R_{\eta_p}(\tau) = (-1)^p r(2B)^{2p} \sum_{m=-\infty}^{\infty} R_\xi(\frac{m}{2W}) d_{2p}(2B\tau - rm). \quad (5.46)$$

Since $\eta_p(t)$ is zero mean, the interpolation noise is wide sense stationary. The corresponding interpolation noise variance is

$$\overline{\eta_p^2} = R_{\eta_p}(0)$$

$$= (-1)^p r(2B)^{2p} \sum_{m=-\infty}^{\infty} R_\xi(\frac{m}{2W}) d_{2p}(rm). \quad (5.47)$$

A spectral density description of the process can be obtained by first transforming (5.46)

$$R_{\eta_p}(t) \leftrightarrow S_{\eta_p}(u)$$

$$= \frac{(2\pi u)^{2p}}{2W} \sum_{m=-\infty}^{\infty} R_\xi(\frac{m}{2W}) e^{-j\pi mu/W} \Pi(\frac{u}{2B}).$$

Application of the Poisson sum formula to the m sum gives

$$S_{\eta_p}(u) = (2\pi u)^{2p} \sum_{n=-\infty}^{\infty} S_\xi(u - 2nW) \Pi(\frac{u}{2B})$$

where $R_\xi(t) \leftrightarrow S_\xi(u)$ and $S_\xi(u)$ is the input noise power spectral density.

An alternate expression for the output noise level follows as

$$\overline{\eta_p^2} = \int_{-\infty}^{\infty} S_{\eta_p}(u) du$$

$$= (2\pi)^{2p} \sum_{n=-\infty}^{\infty} \int_{-B}^{B} u^{2p} S_\xi(u - 2nW) \, du. \quad (5.48)$$

For the unfiltered case ($r = 1$), the integration interval in (5.48) is over $| u | < W$. Since $S_\xi(u) \geq 0$, filtering always results in a noise level equal to or better than the unfiltered case.

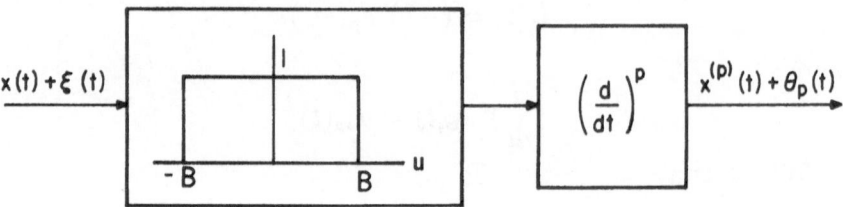

Figure 5.10: Cascaded low-pass filter and p^{th} order differentiator. The output noise level, $\overline{\theta_p^2}$, is a lower bound
for p^{th} order derivative interpolation from the input samples.

5.1.4.1   A Lower Bound on the NINV

Consider $\xi(t)$ input into the cascaded low-pass filter and p^{th}-
order differentiator in Fig. 5.10. Let $\theta_p(t)$ denote the output.
Recall that, in general, the output spectral density $S_\theta(u)$ due
to a spectral density input $S_i(u)$ into a system with transfer
function $H(u)$ is:

$$S_\theta(u) = |H(u)|^2 S_i(u).$$

Thus

$$S_\theta(u) = (2\pi u)^{2p} S_\xi(u) \, \Pi(\frac{u}{2B})$$

and

$$\overline{\theta_p^2} = (2\pi)^{2p} \int_{-B}^{B} u^{2p} S_\xi(u) du. \tag{5.49}$$

Compare this to (5.48). Since $S_\xi(u) \geq 0$, it follows that $\overline{\theta_p^2}$ is a
lower bound for the output noise level

$$\overline{\eta_p^2} \geq \overline{\theta_p^2}.$$

Equality is achieved when $\xi(t)$ has band-limited spectral den-
sity (say over the interval $|u| < \Omega$) and the sampling rate is

sufficiently high to avoid aliasing (*i.e.* $2W - \Omega > B$). A lower sampling rate would result in aliasing and a higher output noise level.

For finite $\overline{\xi^2}$, $S_\xi(u) \to 0$ as $| u | \to \infty$. We see from (5.48) that $\overline{\eta_p^2} \to \overline{\theta_p^2}$ as $2W \to \infty$. Hence, the bound can be approached arbitrarily closely by an appropriate increase in sampling rate. Note that we can guarantee from (5.48) that $\overline{\eta_p^2}$ strictly decreases with r if $S_\xi(u)$ strictly decreases with $u > 0$. This spectral density property is applicable to a Laplace autocorrelation. It is, however, not applicable to triangular autocorrelation.

5.1.4.2 Examples

(a) Triangular Autocorrelation

Consider the triangle autocorrelation parameterized by $a > 0$.

$$R_\xi(\tau) = \overline{\xi^2} \Lambda(\frac{\tau}{a}). \tag{5.50}$$

Substituting into (5.47) gives the normalized error.

$$\frac{\overline{\eta_p^2}}{\overline{\xi^2}} = (-1)^p r(2B)^{2p}[d_{2p}(0) + 2 \sum_{m=1}^{N} (1 - \frac{m}{T})d_{2p}(rm)] \tag{5.51}$$

where

$$T = 2Wa$$

and N is the greatest integer not exceeding T. Plots of (5.51) are shown in Fig. 5.11 for $2B = 1$ and $a = 0.5$ and 0.1.

Of specific interest is the case where sampling is performed such that $T < 1$. The noise samples are then white. That is,

$$R_\xi(\frac{n}{2W}) = \overline{\xi^2}\, \delta[n]. \tag{5.52}$$

Since

$$\begin{aligned} d_{2p}(0) &= \int_{\frac{-1}{2}}^{\frac{1}{2}} (j2\pi u)^{2p}\, du \\ &= \frac{(-1)^p \pi^{2p}}{2p + 1}, \end{aligned} \tag{5.53}$$

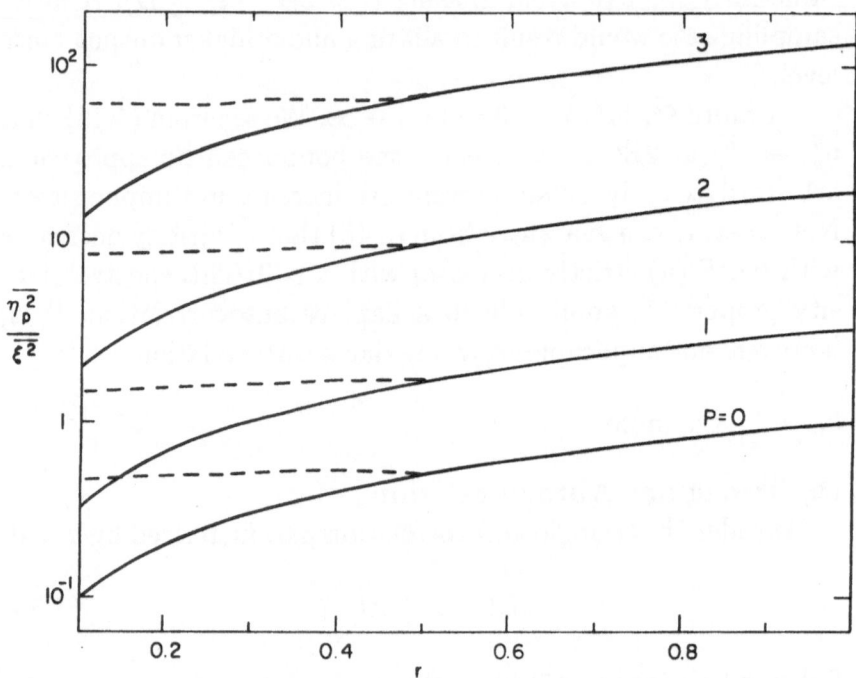

Figure 5.11: NINV for the triangle autocorrelation with $2B = 1$. The solid curve is for $a = 0.1$ and the dashed curve for $a = 0.5$. The curves are identical for $r > 1/2$ where the noise samples are white. For $a = 0.1$, the noise samples are white for $r > 0.1$.

we have for white samples

$$\overline{\eta_p^2}/\overline{\xi^2} = \frac{r(2\pi B)^{2p}}{2p+1}. \tag{5.54}$$

For the conventional sampling theorem, $p = 0$ and the noise level is improved by a factor of r. This result is also in (5.11).

Since $2B = 1$ in Fig. 5.11, the plots there are equivalent to (5.54) for $r > a$. Note that (5.54) is independent of the a parameter - thus the merging of the $a = 0.1$ and 0.5 plots at $r = 0.5$. For the domain shown, all of the $a = 0.1$ samples are white.

Note also that (a) the NINV increases dramatically with the order of differentiation and (b) there can exist a point whereupon a further increase of sampling rate results in an insignificant improvement in the interpolation noise level.

Since $\Lambda(t) \leftrightarrow \mathrm{sinc}^2 u$, from (5.49), the normalized lower bound for the triangle autocorrelation is

$$\overline{\theta_p^2}/\overline{\xi^2} = 2a(2\pi)^{2p} \int_0^B u^{2p} \, \mathrm{sinc}^2(au)\,du. \tag{5.55}$$

To place this in more palatable form, we rewrite it as

$$\overline{\theta_p^2}/\overline{\xi^2} = \frac{(2\pi)^{2p}}{\pi^2 a} \int_0^B u^{2p-2}[1 - \cos(2\pi a u)]\,du. \tag{5.56}$$

For even index

$$\begin{aligned}
d_{2q}(t) &= \int_{-\frac{1}{2}}^{\frac{1}{2}} (j2\pi\nu)^{2q} e^{-j2\pi\nu t}\,d\nu \\
&= 2(2\pi)^{2q}(-1)^q \int_0^{\frac{1}{2}} \nu^{2q} \cos(2\pi\nu t)\,d\nu.
\end{aligned}$$

For $u = 2W\nu$, it follows from (5.56) that for $p > 1$

$$\overline{\theta_p^2}/\overline{\xi^2} = \frac{2(2B)^{2p}(-1)^p}{a}[d_{2p-2}(0) - d_{2p-2}(2aB)].$$

The $p = 0$ case follows immediately from (5.55) using integration by parts. Thus, using (5.53),

$$\overline{\theta_p^2}/\overline{\xi^2} = \begin{cases} \frac{4^p B^{2p-1}}{a}[\frac{\pi^{2p-2}}{2p-1} + (-1)^p d_{2p-2}(2aB)] & ;p > 0 \\[2mm] \frac{2}{\pi}[\mathrm{Si}(2\pi aB) - \sin(\pi aB)\,\mathrm{sinc}(aB)] & ;p = 0. \end{cases}$$

Lower bounds for each of the plots in Fig. 5.11 are graphically indistinguishable from the $r = 0.1$ values.

(b) Laplace Autocorrelation

A second tractable solution from the Laplace correlation parameterized by λ:

$$R_\xi(\tau) = \overline{\xi^2}\, e^{-\lambda|\tau|} \tag{5.57}$$

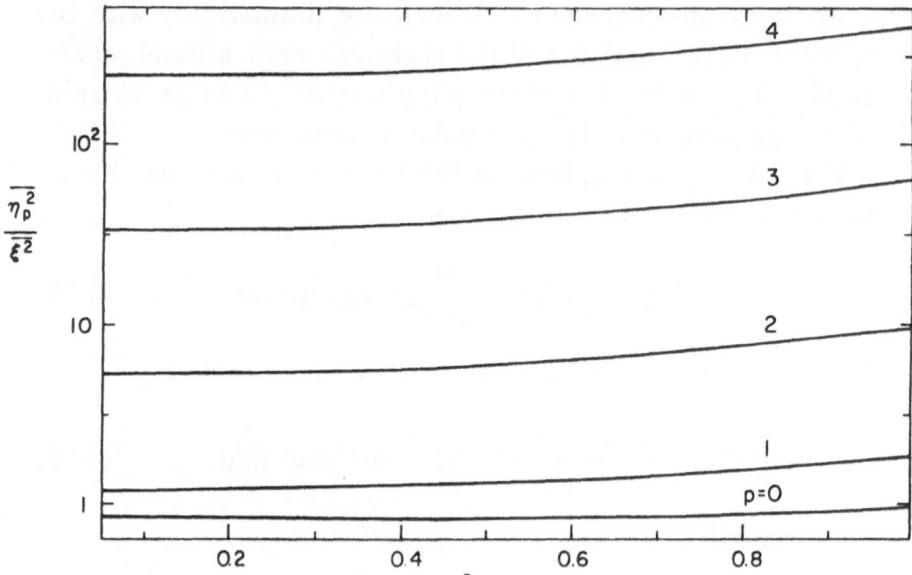

Figure 5.12: NINV for Laplace autocorrelation with
$$\lambda = 2B = 1$$

is considered here. Rewriting (5.47) as

$$\overline{\eta_p^2} = \frac{(2\pi)^{2p}}{2W} \int_{-B}^{B} u^{2p} \sum_{n=-\infty}^{\infty} R_\xi(\frac{n}{2W}) \, e^{-j\pi nu/W} \, du,$$

we can show in a manner similar to that in Example b in Section 5.1.1.2, that the NINV can be written as

$$\overline{\eta_p^2}/\overline{\xi^2} = (2\pi W)^{2p} \sinh(\frac{\lambda}{2W}) \int_0^r \frac{u^{2p} \, du}{\cosh(\frac{\lambda}{2W}) - \cos(\pi u)}. \quad (5.58)$$

The well-behaved (strictly increasing) integrand in (5.58) provides for straightforward digital integration. Sample plots of (5.58) are shown in Fig. 5.12 for $2B = 1$. As with the previous example, the NINV increases significantly with p.

Two special cases of (5.58) are worthy of note.

a) The $p = 0$ case as plotted in Fig. 5.11 simply corresponds to conventional sampling theorem interpolation followed by filtering. For this case, (5.58) becomes the integral in (5.14) whose solution is (5.15).

b) If

$$\frac{\lambda}{2W} \gg 1 \tag{5.59}$$

then (5.58) approaches

$$\begin{aligned}
\overline{\eta_p^2/\xi^2} &= (2\pi W)^{2p} \int_0^r u^{2p} du \\
&= \frac{r(2\pi B)^{2p}}{2p+1} \tag{5.60}
\end{aligned}$$

which is the same result as the discrete white noise case in (5.53). Indeed, when (5.59) is applicable, the noise samples are very nearly white and the plots for white samples in Fig. 5.11 can be used as excellent approximations.

Here, as in the previous example, one must be cautioned on comparing equally parameterized interpolation noise levels for differing p. If we are dealing with temporal functions, the units of $\overline{\eta_p^2/\xi^2}$ are $(\text{seconds})^{-2p}$.

For a lower bound for the Laplace autocorrelation, we transform (5.57) and substitute into (5.49). The result is

$$\overline{\theta_p^2/\xi^2} = \frac{4(2\pi)^{2p}}{\lambda} \int_0^B \frac{\nu^{2p} d\nu}{1 + (\frac{2\pi\nu}{\lambda})^2}. \tag{5.61}$$

Setting $u = 2\pi\nu/\lambda$ gives

$$\overline{\theta_p^2/\xi^2} = \frac{2}{\pi} \lambda^{2p} \int_0^\varepsilon \frac{u^{2p} du}{1 + u^2} \tag{5.62}$$

where $\varepsilon = 2\pi B/\lambda$. The $p = 0$ case follows immediately. For $p > 0$, consider first the case where $\varepsilon < 1$. With $z = -u^2$, the denominator in (5.62) can be expanded via the geometric series:

$$\sum_{m=0}^{\infty} z^m = (1 - z)^{-1}; |z| < 1. \tag{5.63}$$

The resulting integral is evaluated to give

$$\overline{\theta_p^2}/\overline{\xi^2} = \frac{2}{\pi} \lambda^{2p} \sum_{m=0}^{\infty} (-1)^m \frac{\varepsilon^{2m+2p+1}}{2m+2p+1}.$$

Set $n = m + p$ and recall the Taylor series

$$\arctan z = \sum_{n=0}^{\infty} \frac{(-1)^n z^{2n+1}}{2n+1}. \tag{5.64}$$

Thus, for $\varepsilon < 1$,

$$\overline{\theta_p^2}/\overline{\xi^2} = \frac{2}{\pi}(-1)^p \lambda^{2p}[\arctan \varepsilon - \sum_{n=0}^{p-1}(-1)^n \frac{\varepsilon^{2n+1}}{2n+1}].$$

For $\varepsilon > 1$ we rewrite (5.62) as

$$\overline{\theta_p^2}/\overline{\xi^2} = \frac{2}{\pi}\lambda^{2p} [\int_0^{1-} + \int_{1+}^{\varepsilon}] \frac{u^{2p} du}{1+u^2}$$

$$= \frac{2}{\pi}\lambda^{2p}[(-1)^p\{\frac{\pi}{4} - \sum_{n=0}^{p-1} \frac{(-1)^n}{2n+1}\} + \int_{1+}^{\varepsilon} \frac{u^{2p} du}{1+u^2}].$$

From (5.63), it follows that

$$-\sum_{n=0}^{\infty} z^{-n} = \frac{z}{1-z} \qquad ; | z | > 1.$$

Again, with $z = -u^2$, we obtain

$$\overline{\theta_p^2}/\overline{\xi^2} = \frac{2}{\pi}\lambda^{2p}[(-1)^p\{\frac{\pi}{4} - \sum_{n=0}^{p-1} \frac{(-1)^n}{2n+1}\} + \sum_{m=0}^{\infty}(-1)^m \frac{\varepsilon^{2p-2m-1}-1}{2p-2m-1}].$$

Set $n = m - p$ in the m sum and use (5.64). Recognizing $\pi/2 - \arctan(1/\varepsilon) = \arctan \varepsilon$ for $\varepsilon > 0$ again yields (5.65). Placing these results in recursive form gives

$$\frac{\overline{\theta_p^2}}{\overline{\xi^2}} = \begin{cases} \frac{2}{\pi} \arctan \varepsilon & ; p = 0 \\[2mm] \frac{2\lambda^2}{\pi}[\varepsilon - \arctan \varepsilon] & ; p = 1 \\[2mm] \frac{2\lambda(2\pi B)^{2p-1}}{\pi(2p-1)} - \lambda^2 \frac{\overline{\theta_{p-1}^2}}{\overline{\xi^2}} & ; p > 1 \end{cases} \tag{5.65}$$

where
$$\varepsilon = \frac{2\pi B}{\lambda}.$$

Graphically, these bounds are also indistinguishable from the corresponding smallest values on the plots in Fig. 5.12.

5.2 Jitter

Jitter occurs when samples are taken near to but not exactly at the desired sample locations. We here consider only the case of direct uniform sampling [Papoulis (1966)]. Instead of the sample set $x(n/2W)$, we have the sample set

$$\{x(\frac{n}{2W} - \sigma_n) \mid -\infty < n < \infty\}$$

where σ_n is the jitter offset of the nth sample. We assume that the jitter offsets are unknown. If they are known, then sampling theorems for irregularly spaced samples can be used [Marvasti].

In this section, we will show that cardinal series interpolation of jittered samples yields a biased estimate of the original signal. Although the bias can be corrected with an inverse filter, the resulting interpolation noise variance is increased.

5.2.1 Filtered Cardinal Series Interpolation

We know that

$$x(t) = r \sum_{n=-\infty}^{\infty} x(\tfrac{n}{2W}) \operatorname{sinc}(2Bt - rn)$$

and thus may be motivated to estimate $x(t)$ via

$$y(t) = r \sum_{n=-\infty}^{\infty} x(\tfrac{n}{2W} - \sigma_n) \operatorname{sinc}(2Bt - rn). \tag{5.66}$$

The interpolation error is obtained by subtracting these two expressions:

$$\eta(t) = y(t) - x(t)$$
$$= r \sum_{n=-\infty}^{\infty} \xi_n \operatorname{sinc}(2Bt - rn) \tag{5.67}$$

where

$$\xi_n = x\left(\tfrac{n}{2W} - \sigma_n\right) - x\left(\tfrac{n}{2W}\right). \tag{5.68}$$

If the jitter deviations $\{\sigma_n\}$ are identically distributed random variables, then the interpolation in (5.66), does not give an unbiased estimate of $x(t)$. That is

$$\overline{y(t)} = \widetilde{x(t)} \neq x(t) \tag{5.69}$$

where, for any function $v(t)$, we define

$$\widetilde{v(t)} = v(t) * f_\sigma(t) \tag{5.70}$$

and $f_\sigma(t)$ is the probability density function describing each σ_n. To show this, we first note that if $x(t)$ is bandlimited with bandwidth B, then so is $\widetilde{x(t)}$. Hence

$$\widetilde{x(t)} = r \sum_{n=-\infty}^{\infty} x\left(\tfrac{n}{2W}\right) \operatorname{sinc}(2Bt - rn).$$

The expectation of a single jittered sample is

$$\overline{x\left(\tfrac{n}{2W} - \sigma_n\right)} = \int_{-\infty}^{\infty} f_\sigma(\tau)\, x\left(\tfrac{n}{2W} - \tau\right) d\tau$$

$$= \widetilde{x\left(\tfrac{n}{2W}\right)}. \tag{5.71}$$

Substituting this into the expected version of (5.66) substantiates our claim in (5.69).

5.2.2   Unbiased Interpolation from Jittered Samples

Equation (5.69) reveals that cardinal series interpolation from jittered samples results in a biased estimate of the original signal. Motivated by our analysis in the previous section and Section 4.1.2.2., we propose the interpolation formula

$$z(t) = \sum_{n=-\infty}^{\infty} x\left(\frac{n}{2W} - \sigma_n\right) k\left(t - \frac{n}{2W}\right) \tag{5.72}$$

where

$$k(t) \longleftrightarrow \frac{1}{2W} \frac{\Pi\left(\tfrac{u}{2B}\right)}{\Phi_\sigma(u)} \tag{5.73}$$

and the jitter offset's characteristic function, $\Phi_\sigma(u)$, is the Fourier transform of its probability density function:

$$f_\sigma(t) \longleftrightarrow \Phi_\sigma(u).$$

We claim that $z(t)$ is an unbiased estimate of $x(t)$:

$$\overline{z(t)} = x(t). \tag{5.74}$$

Furthermore, the interpolation noise variance is given by

$$\text{var } z(t) = E\left[\,\{z(t) - \overline{z(t)}\}^2\,\right]$$
$$= \sum_{n=-\infty}^{\infty} [\,x^2(\widetilde{\frac{n}{2W}}) - x(\widetilde{\frac{n}{2W}})^2\,]\, k^2(t - \frac{n}{2W}) \tag{5.75}$$

when the jitter offsets are independent. As jitter becomes less and less pronounced,

$$f_\sigma(t) \to \delta(t).$$

From (5.71), we thus expect to see the interpolation noise variance in (5.75) correspondingly approach zero.

Proof: Expectating (5.72) and substituting (5.71) gives

$$\overline{z(t)} = \sum_{n=-\infty}^{\infty} x(\widetilde{\frac{n}{2W}}) k(t - \frac{n}{2W}).$$

Fourier transforming and using (5.73) gives

$$\overline{z(t)} \leftrightarrow \frac{1}{2W} \sum_{n=-\infty}^{\infty} x(\widetilde{\frac{n}{2W}})\, e^{-j\pi n u/W}\, \frac{\Pi(\frac{u}{2B})}{\Phi_\sigma(u)}$$
$$= [X(u)\Phi_\sigma(u)]\, \frac{\Pi(\frac{u}{2B})}{\Phi_\sigma(u)}$$
$$= X(u) \tag{5.76}$$

where we have recognized from (5.70) that

$$\widetilde{x(t)} \leftrightarrow X(u)\, \Phi_\sigma(u).$$

Inverse transforming (5.76) completes our proof of (5.74).

To show (5.75), we first compute the autocorrelation

$$R_z(t;\tau) = \sum_{n=-\infty}^{\infty} \sum_{m=-\infty}^{\infty} \overline{x(\tfrac{n}{2W} - \sigma_n)x(\tfrac{m}{2W} - \sigma_m)}k(t - \tfrac{n}{2W})k(\tau - \tfrac{m}{2W}).$$

$$(5.77)$$

Since the σ_n's are independent

$$\overline{x(\tfrac{n}{2W} - \sigma_n)x(\tfrac{m}{2W} - \sigma_m)} = \begin{cases} \overline{x^2(\tfrac{n}{2W})} & ;n = m \\[2ex] \overline{x(\tfrac{n}{2W})}\,\overline{x(\tfrac{m}{2W})} & ;n \neq m. \end{cases}$$

Substituting into (5.77) gives

$$\begin{aligned} R_z(t;\tau) &= \sum_{n=-\infty}^{\infty} \overline{x^2(\tfrac{n}{2W})}k(t - \tfrac{n}{2W})k(\tau - \tfrac{n}{2W}) \\ &\quad + \sum_{n=-\infty}^{\infty}\sum_{m=-\infty}^{\infty} \overline{x(\tfrac{n}{2W})}\,\overline{x(\tfrac{m}{2W})}k(t - \tfrac{n}{2W})k(\tau - \tfrac{m}{2W}) \\ &\quad - \sum_{n=-\infty}^{\infty} \overline{x(\tfrac{n}{2W})}^2 k(t - \tfrac{n}{2W})k(\tau - \tfrac{n}{2W}) \\ &= \sum_{n=-\infty}^{\infty} [\overline{x^2(\tfrac{n}{2W})} - \overline{x(\tfrac{n}{2W})}^2]k(t - \tfrac{n}{2W})k(\tau - \tfrac{n}{2W}) + \overline{x(t)x(\tau)} \end{aligned}$$

Using the relationship

$$\operatorname{var} z(t) = R_z(t;t) - \overline{x^2(t)}$$

gives our desired result.

5.2.3   In Stochastic Bandlimited Signal Interpolation

Our analysis in this section will show that the use of an inverse filter to obtain an unbiased interpolated estimate from jittered samples will increase the variance of the estimate. Instead of a deterministic signal, $x(t)$, we will consider the analysis of the previous section as applied to a wide sense stationary stochastic signal, $\chi(t)$, with mean $\overline{\chi}$ and autocorrelation $R_\chi(\tau)$.

We will assume that jitter locations (which we will express in vector form as $\vec{\sigma}$) are independent of $\chi(t)$. Thus, the joint probability density function for $\vec{\sigma}$ and $\chi(t)$ can be expressed as the product of the probablility density of $\vec{\sigma}$ with that of $\chi(t)$. Thus, the expectation of any function $w[\vec{\sigma}; \chi(t)]$ can be written

$$E w[\vec{\sigma}; \chi(t)] = E_\chi \, E_{\vec{\sigma}} w[\vec{\sigma}; \chi(t)] \qquad (5.78)$$

where E_χ and $E_{\vec{\sigma}}$ denote expectation with respect to $\chi(t)$ and $\vec{\sigma}$ respectively. Thus, if (5.72) is used to interpolate jittered samples from the stochastic process $\chi(t)$, we conclude from (5.74) that

$$E_{\vec{\sigma}} z(t) = \chi(t).$$

In accordance with (5.78) we expectate both sides with respect to $\chi(t)$ and conclude that

$$E z(t) = \overline{\chi(t)}. \qquad (5.79)$$

If $v(t)$ is any stochastic signal with constant expectation $\overline{v}$, we conclude from (5.70) that

$$
\begin{aligned}
E\widetilde{v(t)} &= \int_{-\infty}^{\infty} \overline{v(\tau)} f_\sigma(t - \tau) d\tau \\
&= \overline{v}.
\end{aligned}
$$

Reinterpret (5.75) as

$$E_{\vec{\sigma}} \, [\, z(t) - \overline{z(t)}\,]^2 = \sum_{n=-\infty}^{\infty} [\chi^2(\widetilde{\tfrac{n}{2W}}) - \chi(\widetilde{\tfrac{n}{2W}})^2] \, k^2(t - \tfrac{n}{2W}).$$

Expectation of both sides with respect to χ therefore gives

$$\operatorname{var} z(t) = \operatorname{var}(\chi(t)) \sum_{n=-\infty}^{\infty} k^2(t - \tfrac{n}{2W})$$

where

$$\operatorname{var} \chi(t) = \overline{\chi^2} - \overline{\chi}^2$$

is a constant. Applying the Poisson sum formula gives

$$\operatorname{var} z(t) = 2W \operatorname{var} [\chi(t)] \sum_{n=-\infty}^{\infty} \wp(2nW) \, e^{-j4\pi nWt}$$

where
$$\wp(u) = K(u) \star K(u).$$

Since $K(u) = 0$ for $|u| > B$, we conclude that $\wp(u) = 0$ for $|u| > 2B$. Thus
$$\wp(2nW) = \wp(0)\delta[n].$$

Since
$$\wp(0) = \frac{1}{(2W)^2} \int_{-B}^{B} |\Phi_\sigma(u)|^{-2} \, du$$

we conclude that

$$\frac{\operatorname{var} z(t)}{\operatorname{var} \chi(t)} = \frac{1}{2W} \int_{-B}^{B} |\Phi_\sigma(u)|^{-2} \, du. \tag{5.80}$$

This and equation (5.79) are our desired results.

5.2.3.1   NINV of Unbiased Restoration

Using the filtered cardinal series in (5.66) on the stochastic signal $\chi(t)$ results in a now unbiased estimate

$$\overline{y(t)} = \overline{\chi}$$

where, now

$$\frac{\operatorname{var} y(t)}{\operatorname{var} \chi(t)} = r. \tag{5.81}$$

Since density functions are non-negative, the characteristic function obeys the inequality

$$|\Phi_\sigma(u)| \le \Phi_\sigma(0) = 1. \tag{5.82}$$

Thus, (5.80) always equals or exceeds r. The price of an unbiased estimate is a higher NINV. This price can be measured in the ratio of (5.80) to (5.81):

$$\rho = \frac{\operatorname{var} z(t)}{\operatorname{var} y(t)} = 2 \int_0^{1/2} |\Phi_\sigma(2Bv)|^{-2} \, dv \ge 1.$$

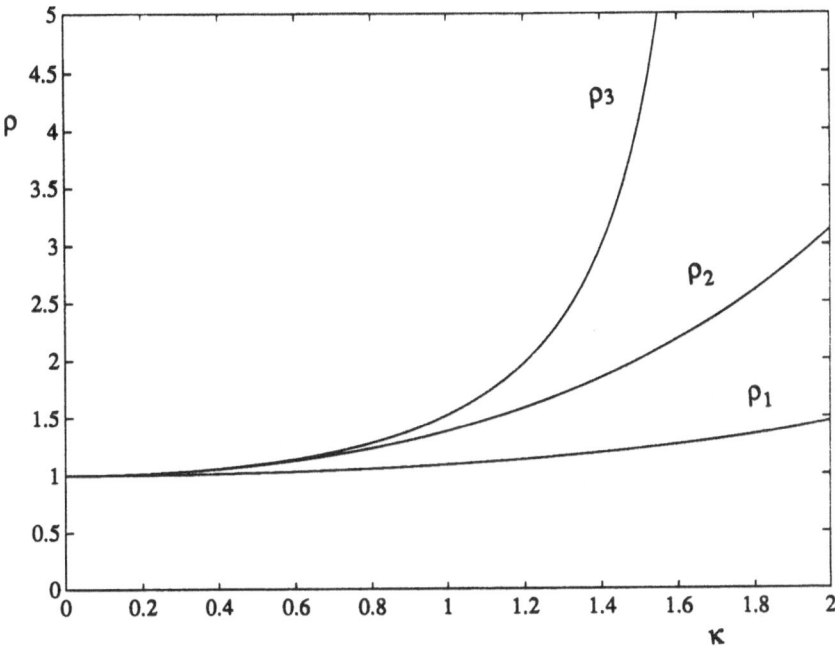

Figure 5.13: A plot of ρ for Gaussian (ρ_1), Laplace (ρ_2) and uniform (ρ_3) jitter as a function of $\kappa = 2\pi Bs$.

5.2.3.2   Examples

In the following specific examples of unbiased restoration from jittered samples, s is the standard deviation of the jitter density. We will also find useful the parameter

$$\kappa = 2\pi Bs.$$

In all cases, we will see ρ increase with increasing κ.

1 Gaussian Jitter

If

$$f_\sigma(t) = \frac{1}{\sqrt{2\pi}s} \exp(\frac{-t^2}{2s^2})$$

then

$$\Phi_\sigma(u) = \exp[-2(\pi su)^2].$$

Thus

$$\rho = 2 \int_0^{1/2} e^{(2\kappa v)^2} dv.$$

A plot is shown in Fig. 5.13 versus κ.

2 Laplace Jitter

Let

$$f_\sigma(t) = \frac{1}{\sqrt{2}s} \exp(\frac{-\sqrt{2}\,|t|}{s}).$$

Then

$$\Phi_\sigma(u) = \frac{2/s^2}{(2/s^2) + (2\pi u)^2}$$

and we obtain the closed form solution

$$\rho = 1 + \frac{1}{3}\kappa^2 + \frac{1}{2\sigma}\kappa^4$$

a plot is shown in Fig. 5.13.

3 Uniform Jitter

Uniform jitter is characterized by the density

$$f_\sigma(t) = \frac{1}{\varphi} \, \Pi(t/\varphi) \, \leftrightarrow \, \text{sinc}(\varphi u)$$

where $\varphi = \sqrt{12}s$. The density function is thus zero for

$$|t| > \frac{\varphi}{2} = \frac{\sqrt{12}\,s}{2}.$$

Therefore

$$\rho = 2 \int_0^{1/2} \frac{dv}{\text{sinc}^2(\frac{\sqrt{12}\,\kappa v}{\pi})}.$$

A plot is shown in Fig. 5.13. Note that there is a singularity in the integrand when a zero of the sinc lies within the interval of integration. This occurs when

$$\varphi \geq \frac{1}{B}$$

or, in other words, when the temporal locations of two adjacent samples have a finite probability of interchanging. The value of ρ in such cases is unbounded. Obtaining an unbiased estimate by inverse filtering is therefore an unstable undertaking.

5.3 Truncation Error

The cardinal series requires an infinite number of terms to exactly interpolate a bandlimited signal from its samples. In practice, only a finite number of terms can be used. We will examine the resulting *truncation error* for both deterministic and stochastic bandlimited signals.

5.3.1 An Error Bound

We can write the truncated cardinal seies approximation of $x(t)$ as

$$x_N(t) = \sum_{n=-N}^{N} x(\tfrac{n}{2B}) \operatorname{sinc}(2Bt - n).$$

The error resulting from using a finite number of terms is referred to as truncation error:

$$e_N(t) = \mid x(t) - x_N(t) \mid^2 . \tag{5.83}$$

We will show that [Papoulis (1966)]

$$
\begin{aligned}
e_N(t) \;\leq\; & 2B(E - E_N)[\operatorname{sinc}(2Bt + N) - \operatorname{sinc}(2Bt - N)](-1)^n \frac{\sin(2\pi Bt)}{\pi} \\
=\;\; & (E - E_N)\frac{4NB/\pi^2}{N^2 - (2Bt)^2}\sin^2(2\pi Bt) \;\; ; \mid t \mid < \frac{N}{2B} \tag{5.84}
\end{aligned}
$$

where E is the energy of $x(t)$ and E_N is the energy of $x_N(t)$. Using Parseval's theorem in (3.32) applied to $x_N(t)$ gives,

$$E_N = \frac{1}{2B} \sum_{n=-N}^{N} \mid x(\tfrac{n}{2B}) \mid^2 .$$

From (5.84), as we would expect,

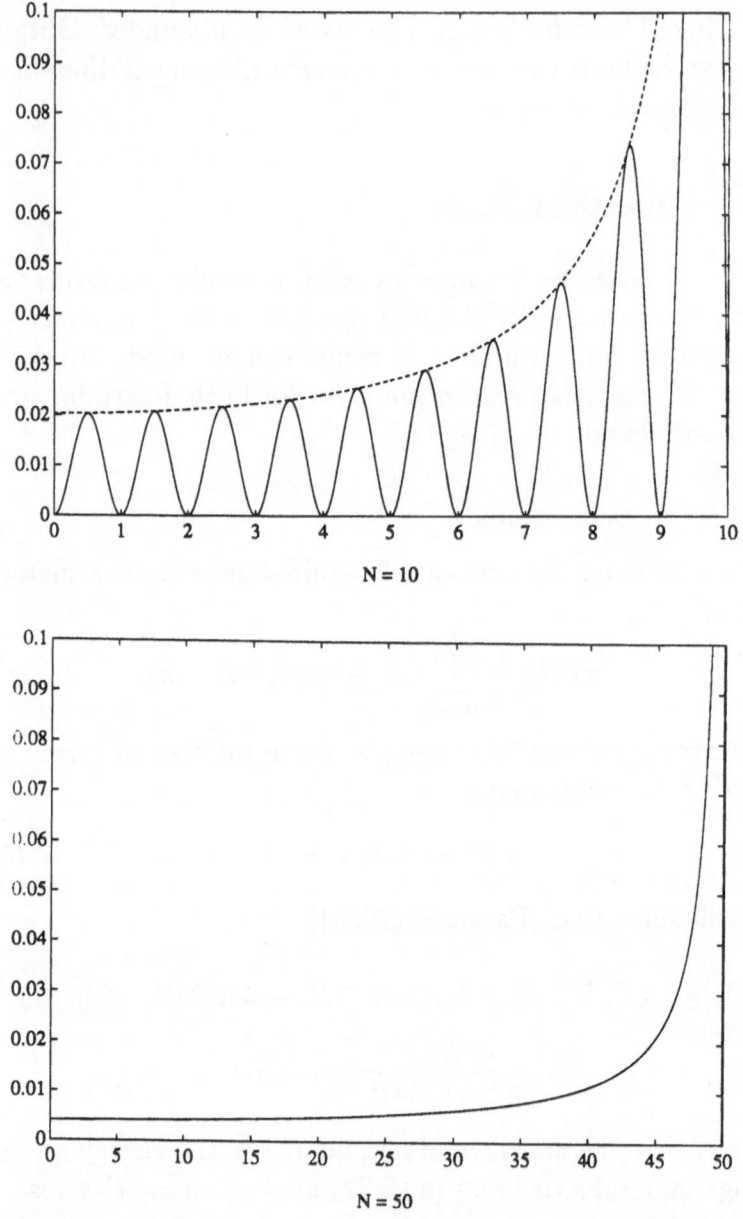

Figure 5.14: Plots of the errors bound in (5.84) as $e_N(\tau/2B)/\{2B(E - E_N)\}$ for $N = 10$ and the envelope of the error bound for $N = 50$.

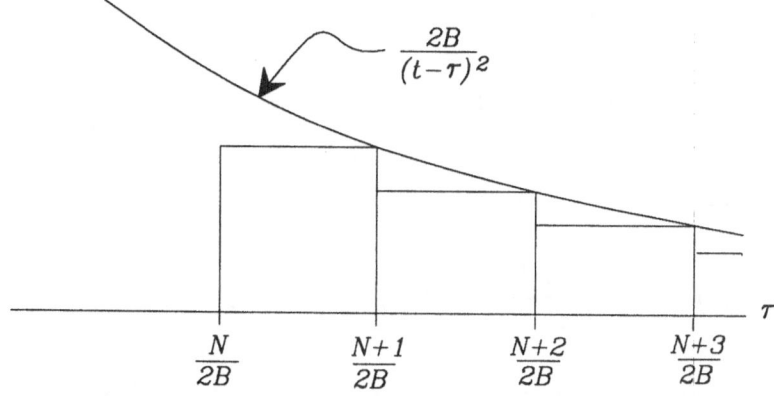

Figure 5.15: The left side of the inequality in (5.87) is the area under the rectangles. The right side of the inequality is the area under the $2B(t - \tau)^{-2}$ curve over the interval $(N/2B, \infty)$.

a) since $E_\infty = E$, the error bound tends to zero as $N \to \infty$.

b) for $\mid n \mid \leq N$, the error $e_N(\frac{n}{2B})$ is zero.

A plot of $e_N(\tau/2B)/\{2B(E - E_N)\}$ is shown at the top of Fig. 5.14. Also shown is the envelope of the curve which is given by the expression

$$\frac{2\,N/\pi^2}{N^2 - \tau^2}.$$

The envelope for $N = 50$ is shown at the bottom of Fig. 5.14. Note that, at $\tau = N - 1$, the envelope has a value, for large N, of about $1/\pi^2 \approx 0.10$.

Proof: Using the cardinal series expansion in (3.4), we have

$$e_N(t) = \frac{1}{\pi^2}\sin^2(2\pi Bt) \mid \sum_{|n|>N} \frac{(-1)^n x(\frac{n}{2B})}{2Bt - n} \mid^2$$

applying Schwarz's inequality gives

$$e_N(t) \leq \frac{\sin^2(2\pi Bt)}{2B\pi^2}(E - E_N) \sum_{|n|>N} (t - \tfrac{n}{2B})^{-2} \qquad (5.85)$$

where we have recognized that

$$\frac{1}{2B} \sum_{|n|>N} |x(\tfrac{n}{2B})|^2 = E - E_N.$$

We can write (5.85) as

$$e_N(t) \leq \frac{\sin^2(2\pi Bt)}{2B\pi^2}(E - E_N)\left[\sum_{n=-\infty}^{-N-1} + \sum_{n=N+1}^{\infty}\right](t - \tfrac{n}{2B})^{-2}. \quad (5.86)$$

Motivated by Fig. 5.15, we can write, for $t < N/2B$,

$$\sum_{n=N+1}^{\infty} (t - \tfrac{n}{2B})^{-2} \leq 2B \int_{N/2B}^{\infty} (t - \tau)^{-2}\, d\tau$$

$$= 2B\left(\frac{N}{2B} - t\right)^{-1}. \qquad (5.87)$$

Similarly, for $t > N/2B$,

$$\sum_{n=N+1}^{\infty} (t + \tfrac{n}{2B})^{-2} \leq 2B\left(\frac{N}{2B} + t\right)^{-1}.$$

Substitution of these two inequalities into (5.86) followed by simplification yields our desired result in (5.84).

5.3.2   Noisy Stochastic Signals

In this section, we invesigate truncation error for a noisy stochastic signal at a sample point [Radbel & Marks] and compare the certainty of the resulting estimate to that obtained from

- Simply using the noisy sample and

- A minimum mean square estimate.

Let $f(t)$ denote a bandlimited wide sense stationary stochastic process with bandwidth B as described in Section 3.5. The filtered cardinal series

$$f(t) = r \sum_{n=-\infty}^{\infty} f(\tfrac{n}{2W}) \operatorname{sinc}(2Bt - rn) \qquad (5.88)$$

converges to $f(t)$ in the mean square sense when $r = B/W < 1$. In other words,

$$E\left[|\, \hat{f}(t) - f(t)\,|^2\right] = 0.$$

Let $\xi(t)$ denote a zero mean wide sense stationary stochastic process with autocorrelation $R_\xi(\tau)$. Our noisy waveform is then

$$g(t) = f(t) + \xi(t).$$

We will assume that the signal and the noise are uncorrelated. Thus

$$R_g(\tau) = R_f(\tau) + R_\xi(\tau).$$

Our goal is to estimate $f(0)$ from the set of samples

$$\{g(\tfrac{n}{2W}) \mid -N \leq n \leq N\}$$

using:

(a) the single sample $g(0)$

(b) the truncated cardinal series

(c) a minimum mean square estimate

Each case will be examined separately and contrasted by the comparison of the resulting mean square error.

Case a: The estimate here is

$$f_A(0) = g(0). \qquad (5.89)$$

The resulting error is

$$\varepsilon_A = E\left[|\, f(0) - f_A(0)\,|^2\right] = \overline{\xi^2}.$$

Case b: The truncated cardinal series expression evaluated at $t = 0$ is

$$f_B(0) = r \sum_{n=-N}^{N} g(\tfrac{n}{2W}) \operatorname{sinc}(rn). \tag{5.90}$$

The mean square error of this estimate is given by

$$
\begin{aligned}
\varepsilon_B(N) &= E\left[|\, f(0) - f_B(0) \,|^2\right] \\
&= E\left[|\, f(0) - r \sum_{n=-N}^{N} g(\tfrac{n}{2W}) \operatorname{sinc}(rn) \,|^2\right] \\
&= R_f(0) - 2r \sum_{n=-N}^{N} R_f(\tfrac{n}{2W}) \operatorname{sinc}(rn) \\
&\quad + r^2 \sum_{n=-N}^{N} \sum_{m=-N}^{N} R_g[\tfrac{n-m}{2W}] \operatorname{sinc}(rn) \operatorname{sinc}(rm).
\end{aligned}
$$
$$\tag{5.91}$$

Case c: To find the minimum mean square solution [Papoulis (1984)], we wish to find the set of coefficients $\{h[n] \mid -N \le n \le N\}$ in the expression

$$f_C(0) = \sum_{n=-N}^{N} g(\tfrac{n}{2W}) h[n] \tag{5.92}$$

that minimizes the error

$$\varepsilon_C(N) = E\left[|\, f(0) - f_C(0) \,|^2\right].$$

We substitute the appropriate expressions and set the derivatives of $\varepsilon_C(N)$ with respect to each $h[n]$ to zero. Using this procedure, the optimal $h[n]$'s can be shown to satisfy the simultaneous set of equations

$$R_f(\tfrac{m}{2W}) = \sum_{n=-N}^{N} R_g[\tfrac{m-n}{2W}] h[n]; \quad -N \le m \le N. \tag{5.93}$$

Solving these equations and using (5.92) results in a minimum mean square error of

$$\varepsilon_C(N) = R_f(0) - \sum_{n=-N}^{N} R_f(\tfrac{n}{2W})\, h[n].$$

NOTES

1. The truncated sampling theorem estimate will always be better than the minimum mean square error estimate, *i.e.* $\varepsilon_B(N) \geq \varepsilon_C(N)$. This, however, comes at a price. The minimum mean square estimate is *parametric* in the sense that solution of the equations in (5.93) requires detailed information about the second order statistics of both the signal and the noise. The sampling theorem estimate in (5.90), on the other hand, does not.

2. An example of the relative performance of the three estimates is shown in Fig. 5.16 for discrete white noise with variance $\overline{\xi^2}$. The signals with uniform and triangular spectral densities are

$$S_f(u) = \overline{f^2}\,\frac{\Pi(\tfrac{u}{2B})}{2B} \tag{5.94}$$

and

$$S_f(u) = \overline{f^2}\,\frac{\Lambda(\tfrac{u}{2B})}{B} \tag{5.95}$$

where

$$\overline{f^2} = R_f(0).$$

3. The error in (5.91) can be partitioned into that due to data noise and that due to truncation. Specifically

$$\varepsilon_B(N) = \varepsilon_D(N) + \varepsilon_T(N)$$

where

$$\begin{aligned}
\varepsilon_T(N) \;=\;& R_f(0) - 2r \sum_{n=-N}^{N} R_f(\tfrac{n}{2W})\,\mathrm{sinc}(rn) \\
&+ r^2 \sum_{n=-N}^{N} \sum_{m=-N}^{N} R_f[\frac{n-m}{2W}]\,\mathrm{sinc}(rn)\,\mathrm{sinc}(rm)
\end{aligned}$$

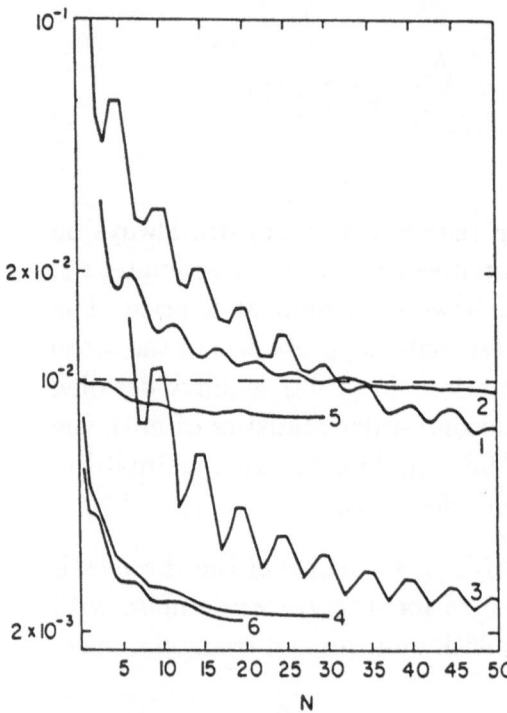

curve	$S_f(u)$	estimate	r
1	u	tst	0.2
2	u	tst	0.8
3	t	tst	0.2
4	u	mmse	0.2
5	u	mmse	0.8
6	t	mmse	0.2

Figure 5.16: The normalized mean square error for truncated estimation of a noisy sample, $f(0)$. In each case, the signal-to-noise ratio is $\overline{f^2}/\overline{\xi^2} = 100$ and the data noise is discrete and white. The sample point estimate in (5.89) has a normalized mean square error of $\varepsilon_A = \overline{\xi^2}/\overline{f^2} = 0.01$. This is shown by the horizontal dashed line.

key:

u =uniform power spectral density as in (5.94).

t =triangular power spectral density as in (5.95).

tst =truncated sampling theorem estimate, $f_B(0)$.

u = mmse= minimum mean square estimate, $f_C(0)$.

and

$$\varepsilon_D(N) = r^2 \sum_{n=-N}^{N} \sum_{m=-N}^{N} R_\xi[\frac{n-m}{2W}] \operatorname{sinc}(rn) \operatorname{sinc}(rm).$$

For a fixed N, $\varepsilon_T(N)$ as a function of r tends to oscillate with a decreasing envelope. The function $\varepsilon_D(N)$, on the other hand, is a strictly increasing function of r. This suggests that there may exist a sampling rate that minimizes the overall mean square error. This is confirmed in Fig. 5.17 where the minimum mean square error, $\varepsilon_B(N)$, is plotted versus r for various N.[1] In those cases where a minimum exists, there exists an optimal sampling rate equal to $2W = 2B/r$.

---

[1]Sometimes, a value lower than N (e.g. $N-1$ or $N-2$) gave a better mean square error. In such cases, the lower mean square error is plotted in Fig. 5.17.

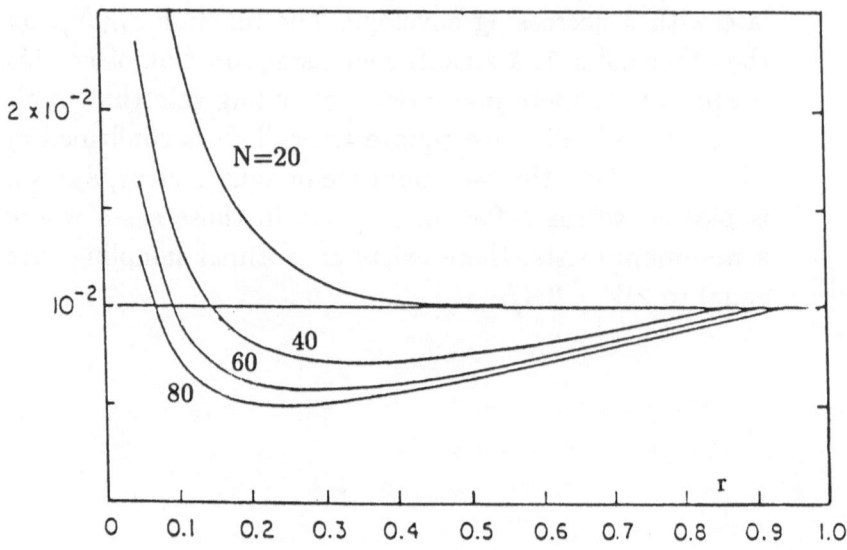

Figure 5.17: A plot of the smallest mean square error that can be achieved with a truncated sampling theorem estimate of $g(0)$ using not more than $2N + 1$ noisy samples. In each case, the signal-to-noise ratio is $\overline{f^2}/\overline{\xi^2} = 100$ and the data noise is discrete and white. The power spectral density of the signal is uniform as in (5.94). Each curve's minimum reflects the optimal sampling rate for the corresponding set of parameters.

5.4 Exercises

5.1 According to (5.11), the filtered interpolation noise level can be made arbitrarily small by simply increasing the sampling rate. Why is this result intuitively not satisfying? Explain why we can not make the noise level arbitrarily small in practice.

5.2 The interpolation function for trigonometric polynomials has infinite energy. Thus, application of either (5.23) or (5.24) results in an infinite NINV classifying the interpolation as ill-posed. Explain why this reasoning is faulty. Derive the filtered and unfiltered NINV using (4.71) assuming additive discrete white noise on the P data points.

5.3 The stochastic process in (5.37) is wide sense *cyclostationary* [Papoulis (1984); Stark & Woods]. Define the stochastic process
$$\Psi(t) = \eta(t - \Theta)$$
where Θ is a uniform random variable on the interval $(0, T_N)$ and $\eta(t)$ is given by (5.37).

(a) Show that $\Psi(t)$ is wide sense stationary.

(b) Evaluate $\overline{\Psi^2}$ in terms of the Fourier coefficients in (5.39).

5.4 In both Figures 5.1 and 5.11, colored noise always produces a worse NINV than white noise. Is this always true? If not, specify a noise for which the statement is not true.

5.5 Using the result of Exercise 2.23, show that the interpolation noise variance due to jitter in (5.67) can be bounded as
$$\overline{\eta^2(t)} \le (2\pi B)^3 E \sigma_{max}/(3\pi)$$
where the jitter deviation is bounded by
$$\sigma_{max} \ge |\sigma_n|$$
and we have assumed sampling at the Nyquist rate.

5.6 Using (5.75), let

$$E_z = \int_{-\infty}^{\infty} \text{var } z(t) \, dt.$$

Similarly, let

$$E_y = \int_{-\infty}^{\infty} \text{var } y(t) \, dt.$$

Show that $E_z \geq E_y$.

5.7 Show that the interpolation noise in (5.67) is not zero mean.

5.8 Compute the NINV in (5.9) when the additive noise has a

Cauchy autocorrelation with parameter γ:

$$R_\xi(\tau) = \frac{\overline{\xi^2}}{(\tau/\gamma)^2 + 1}.$$

5.9 Consider restoring a single lost sample at the origin when $r \leq 1/2$. We could use (4.9) or delete every other sample and use the filtered cardinal series in (4.2). Contrast the corresponding NINV's.

5.10 A bandlimited signal, $x(t)$, with nonzero finite energy, can be characterized by a finite number of samples if there exists a T such that $x(t) \equiv 0$ for $|t| > T$. Show, however, that no such class of signals exists.

5.11 Show that

$$\int_{-\infty}^{\infty} e_N(t) \, dt = E - E_N$$

where $e_N(t)$ is the truncation error in (5.83).

5.12 Show that (5.91) can be massaged into the computationally advantageous iterative form

$$
\begin{aligned}
\varepsilon_B(N+1) \;=\; & \varepsilon_B(N) - 4r R_f[(N+1)/2W] \, \text{sinc}[r(N+1)] \\
& + \; 2r^2 \, \text{sinc}^2[r(N+1)]\{R_g(0) + R_g[(2N+2)/2W]\} \\
& + \; 2r^2 \, \text{sinc}\,[r(N+1)] \sum_{n=-N}^{N} \{R_g[(n+N+1)/2W] \\
& + \; R_g[(n-N-1)/2W]\} \, \text{sinc}(rn)
\end{aligned}
$$

with initialization

$$\varepsilon_B(0) = (1 - r)^2 R_f(0) + R_\xi(0).$$

REFERENCES

R.N. Bracewell, **The Fourier Transform and Its Applications, 2nd edition**, Revised, McGraw-Hill, NY, 1986.

K.F. Cheung and R.J. Marks II "Ill-posed sampling theorems", *IEEE Transactions on Circuits and Systems*, vol. CAS-32, pp.829-835 (1985).

I.S. Gradshteyn and I.M. Ryzhik, **Tables of Integrals, Products and Series**, Academic Press, New York, 4th ed.,1965.

R. W. Hamming, **Numerical Methods for Scientists and Engineers**, McGraw-Hill, New York, 1962.

R.J. Marks II "Noise sensitivity of bandlimited signal derivative interpolation", *IEEE Transactions on Acoustics, Speech and Signal Processing*, vol. ASSP-31, pp.1029-1032 (1983).

R.J. Marks II and D. Radbel "Error in linear estimation of lost samples in an oversampled bandlimited signal", *IEEE Transactions on Acoustics, Speech and Signal Processing*, vol. ASSP-32, pp.648-654 (1984).

F.A. Marvasti, **A Unified Approach to Zero-Crossing and Nonuniform Sampling**, Oak Park, Ill., 1987.

A. Papoulis "Error analysis in sampling theory", *Proc. IEEE*, vol. 54, pp. 947-955 (1966).

A. Papoulis, **Signal Analysis**, McGraw-Hill, New York, 1977.

A. Papoulis, **Probability, Random Variables, and Stochastic Processes**, 2nd Ed., McGraw-Hill, NY, 1984.

D. Radbel and R.J. Marks II "An FIR estimation filter based on the sampling theorem", *IEEE Transactions on Acoustics, Speech and Signal Processing*, vol. ASSP-33, pp.455-460 (1985).

C.E. Shannon "A mathematical theory of communication", *Bell System Technical Journal*, vol. 27, pp.379, 623, (1948).

C.E. Shannon "Communications in the presence of noise", *Proc. IRE*, vol. 37, pp.10-21 (1949).

H. Stark and John W. Woods, **Probability, Random Processes, and Estimation Theory For Engineers**, Prentice-Hall, New Jersey, 1986.

Shiao-Min Tseng, **Noise Level Analysis of Linear Restoration Algorithms for Band-Limited Signals**, Thesis, University of Washington, 1984.

6

The Sampling Theorem in Higher Dimensions

The first generalization of the Shannon sampling theorem to two and more dimensions was done by Peterson and Middleton. Multidimensional signals include black and white images which can be depicted as two dimensional functions with zero corresponding to black and one as white. Intermediate grey levels are classified between these limits.

6.1 Multidimensional Fourier Analysis

A brief review of multidimensional signal representation and Fourier analysis is required prior to our discussion of multidimensional sampling theory. Let

$$\vec{t} = [t_1 \ t_2 \ t_3 \cdots t_N]^T$$

denote an N dimensional vector. (The superscript T denotes transposition.) The function $x(\vec{t})$ assigns a (possibly complex) number over some set of $\vec{t}$'s.

The Fourier transform of $x(\vec{t})$ is defined by the integral

$$X(\vec{u}) = \int_{\vec{t}} x(\vec{t}) \exp(-j2\pi \vec{u}^T \vec{t}) d\vec{t} \qquad (6.1)$$

where

$$\int_{\vec{t}} = \int_{t_1=-\infty}^{\infty} \int_{t_2=-\infty}^{\infty} \cdots \int_{t_N=-\infty}^{\infty}$$

and

$$d\vec{t} = dt_1 \ dt_2 \cdots dt_N$$

1. Transform

$$x(\vec{t}) \leftrightarrow X(\vec{u}) = \int_{\vec{t}} x(\vec{t}) \exp(-j2\pi \vec{u}^T \vec{t}) d\vec{t}$$

2. Linearity

$$\sum_k a_k x_k(\vec{t}) \leftrightarrow \sum_k a_k X_k(\vec{u})$$

3. Inversion

$$x(\vec{t}) = \int_{\vec{u}} X(\vec{u}) \exp(j2\pi \vec{u}^T \vec{t}) d\vec{u} \leftrightarrow X(\vec{u})$$

4. Shift

$$x(\vec{t} - \vec{\tau}) \leftrightarrow X(\vec{u}) \exp(-j2\pi \vec{u}^T \vec{\tau})$$

5. Separability

$$\prod_{n=1}^{N} x_n(t_n) \leftrightarrow \prod_{n=1}^{N} X_n(u_n)$$

6. Rotation and Scale

$$x(\mathbf{A}\vec{t}) \leftrightarrow \frac{X(\mathbf{A}^{-T}\vec{u})}{|\det \mathbf{A}|}$$

7. Convolution

$$x(\vec{t}) * h(\vec{t}) = \int_{\vec{\tau}} x(\vec{\tau}) h(\vec{t} - \vec{\tau}) d\tau \leftrightarrow X(\vec{u}) H(\vec{u})$$

8. Modulation

$$x(\vec{t}) h(\vec{t}) \leftrightarrow X(\vec{u}) * H(\vec{u})$$

Table 6.1: Properties of the multi-dimensional Fourier transform.

6.1.1   Properties

Table 6.1 contains a list of properties of multidimensional Fourier transforms. Many are straightforward generalizations of their one dimensional counterparts. Those that are not warrant further elaboration.

6.1.1.1   Separability

A function $x(\vec{t})$ is said to be separable if it can be written as the product of one dimensional functions:

$$x(\vec{t}) = \prod_{n=1}^{N} x_n(t_n) \tag{6.2}$$

The separability theorem states that the corresponding multidimensional Fourier transform is the product of the one dimensional Fourier transforms of the one dimensional signals:

$$x(\vec{t}) \leftrightarrow X(\vec{u}) = \prod_{n=1}^{N} X_n(u_n)$$

where

$$x_n(t_n) \leftrightarrow X_n(u_n) = \int_{-\infty}^{\infty} x_n(t_n)\exp(-j2\pi u_n t_n)dt_n$$

is a one dimensional transform pair. The proof follows from substituting (6.2) into (6.1) and separating integrals.

Example : The separability theorem allows evaluation of certain multidimensional transforms using one dimensional transforms. Consider the $N = 2$ dimensional example in Figure 6.1 where a function is one inside the two symmetrically- spaced squares and is zero outside. We can write

$$x(t_1,t_2) = \Pi\left(\frac{t_1}{c} - \frac{1}{2}\right)\Pi\left(\frac{t_2}{c} - \frac{1}{2}\right) + \Pi\left(\frac{t_1}{c} + \frac{1}{2}\right)\Pi\left(\frac{t_2}{c} + \frac{1}{2}\right) \tag{6.3}$$

From the separability theorem

$$\Pi\left(\frac{t_1}{c} \pm \frac{1}{2}\right)\Pi\left(\frac{t_2}{c} \pm \frac{1}{2}\right) \leftrightarrow c^2\mathrm{sinc}(cu_1)\mathrm{sinc}(cu_2)e^{\pm j\pi c(u_1+u_2)} \tag{6.4}$$

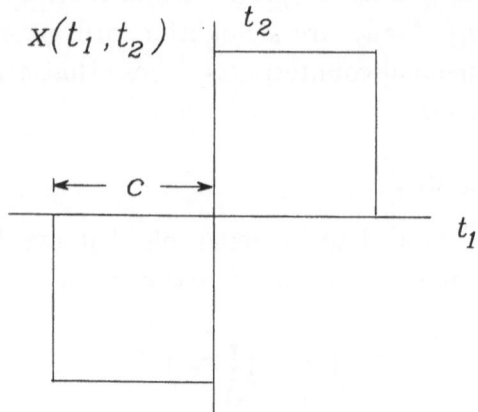

Figure 6.1: A two dimensional signal is one inside the squares and is zero outside.

Superimposing the $+$ and $-$ components of (6.4) and using the linearity property of Fourier transformation gives us the transform of (6.3):

$$x(t_1, t_2) \leftrightarrow 2c^2 \text{sinc}(cu_1)\text{sinc}(cu_2) \cos[\pi c(u_1 + u_2)] \qquad (6.5)$$

The above example is a case where a function is the difference of two separable functions. One can easily establish that, for any two dimensional function, we can find some one dimensional functions such that

$$x(t_1, t_2) = \sum_{m=-\infty}^{\infty} x_m^{(1)}(t_1) x_m^{(2)}(t_2)$$

That is, any two dimensional function can be expressed as the superposition of separable functions. More generally, in N dimensions,

$$x(\vec{t}) = \sum_{m=-\infty}^{\infty} \prod_{n=1}^{N} x_m^{(n)}(t_n). \qquad (6.6)$$

The choices of the one dimensional functions are clearly not

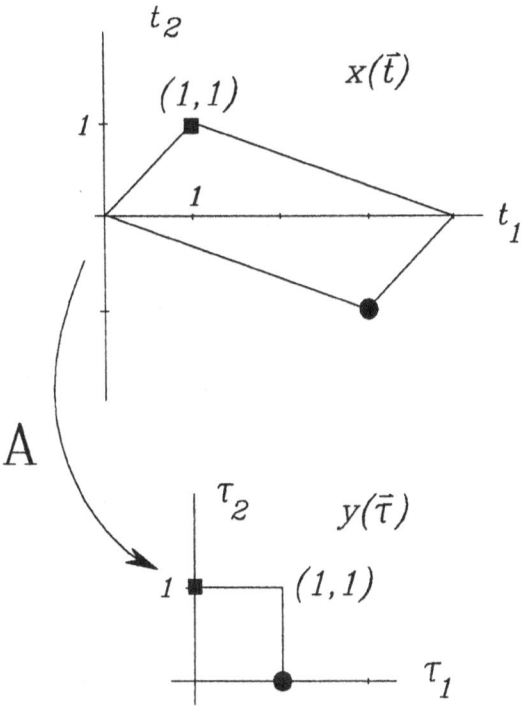

Figure 6.2: Illustration of linear scaling of the unit square (bottom) to a parallelogram (top).

unique. For such a representation, it follows that

$$x(\vec{t}) \leftrightarrow \sum_{m=-\infty}^{\infty} \prod_{n=1}^{N} X_m^{(n)}(u_n). \qquad (6.7)$$

6.1.1.2 Rotation, Scale and Transposition

For any nonsingular matrix $\mathbf{A}$, the function $y(\mathbf{A}\vec{t})$ can be envisioned as a rotation, scaling, and/or transposition of the function $y(\vec{t})$. We illustrate this with the unit square at the bottom of Figure 6.2. What linear coordinate transformation, $\vec{\tau} = \mathbf{A}\vec{t}$, will map this to the shaded parallelogram, $x(\vec{t})$, shown at the top of

Figure 6.2? If we choose to map the corner with the square on the bottom to the corner with the square on the top, we must satisfy

$$\begin{bmatrix} 0 \\ 1 \end{bmatrix} = \mathbf{A} \begin{bmatrix} 1 \\ 1 \end{bmatrix}$$

Similarly, for the circles,

$$\begin{bmatrix} 1 \\ 0 \end{bmatrix} = \mathbf{A} \begin{bmatrix} 3 \\ -1 \end{bmatrix}$$

Combining gives

$$\begin{bmatrix} 1 & 0 \\ 0 & 1 \end{bmatrix} = \mathbf{A} \begin{bmatrix} 3 & 1 \\ -1 & 1 \end{bmatrix}$$

or

$$\mathbf{A} = \begin{bmatrix} 3 & 1 \\ -1 & 1 \end{bmatrix}^{-1} = \frac{1}{4} \begin{bmatrix} 1 & -1 \\ 1 & 3 \end{bmatrix}$$

Thus

$$\vec{\tau} = \begin{bmatrix} \frac{t_1 - t_2}{4} \\ \frac{t_1 - 3t_2}{4} \end{bmatrix}$$

Since

$$y(\vec{\tau}) = \Pi \left(\tau_1 - \frac{1}{2} \right) \Pi \left(\tau_2 - \frac{1}{2} \right)$$

we conclude that

$$x(\vec{t}) = y(\mathbf{A}\vec{t}) = \Pi \left[\frac{1}{4}(t_1 - t_2) - \frac{1}{2} \right] \Pi \left[\frac{1}{4}(t_1 + 3t_2) - \frac{1}{2} \right]$$

which is our desired answer.

6.1.1.2.1   Scale

Consider the rectangular box shown in Figure 6.3. Scaling to the unit square shown at the bottom of Figure 6.2 is performed by $\vec{\tau} = \mathbf{S}\vec{t}$ where

$$\mathbf{S} = \begin{bmatrix} \frac{1}{M_1} & 0 \\ 0 & \frac{1}{M_2} \end{bmatrix} \qquad (6.8)$$

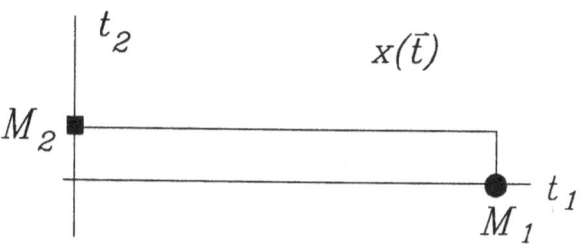

Figure 6.3: Scaling the unit square in Figure 6.2 by M_1 in the t_1 direction and M_2 in the t_2 direction.

Thus

$$x(\vec{t}) = \Pi\left(\frac{t_1}{M_1} - \frac{1}{2}\right)\Pi\left(\frac{t_2}{M_2} - \frac{1}{2}\right)$$

More generally, multidimensional scaling is performed using

$$\mathbf{S} = \mathrm{diag}\left[\frac{1}{M_1}\,\frac{1}{M_2}\cdots\frac{1}{M_N}\right] \tag{6.9}$$

where diag denotes a diagonal matrix.

6.1.1.2.2   Rotation

To rotate a function, $y(\vec{t})$, counterclockwise by an angle of θ, we write $y(\mathbf{R}\vec{t})$ where

$$\mathbf{R} = \begin{bmatrix} \cos(\theta) & \sin(\theta) \\ -\sin(\theta) & \cos(\theta) \end{bmatrix} \tag{6.10}$$

Note that $\mathbf{R}^{-1} = \mathbf{R}^T$. The rotated square in Figure 6.4 can thus be written as

$$x(\vec{t}) = \Pi\left[t_1\cos(\theta) + t_2\sin(\theta) - \frac{1}{2}\right]\Pi\left[t_1\sin(\theta) - t_2\cos(\theta) + \frac{1}{2}\right]$$

If we wish to first rotate and then scale, we scale the rotated function, $y(\mathbf{R}\vec{t})$, to $y(\mathbf{RS}\vec{t})$. For the function in Figure 6.5, for

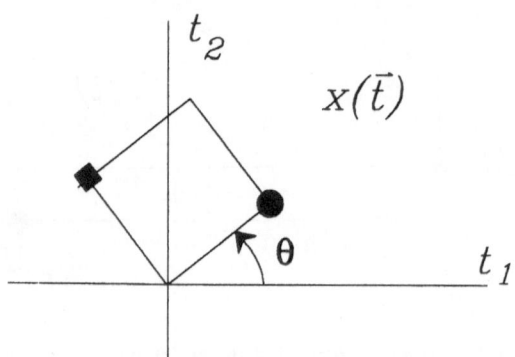

Figure 6.4: Rotation of the unit square in Figure 6.2 counter-clockwise by an angle of θ.

example, we rotate the unit square at the bottom of Figure 6.2 by $\theta = 45°$ and scale in the t_2 direction by a factor of 2. The resulting scaling matrix is

$$\mathbf{RS} = \begin{bmatrix} \frac{1}{\sqrt{2}} & \frac{1}{\sqrt{2}} \\ -\frac{1}{\sqrt{2}} & \frac{1}{\sqrt{2}} \end{bmatrix} \begin{bmatrix} 1 & 0 \\ 0 & \frac{1}{2} \end{bmatrix} = \begin{bmatrix} \frac{1}{\sqrt{2}} & \frac{1}{2\sqrt{2}} \\ -\frac{1}{\sqrt{2}} & \frac{1}{2\sqrt{2}} \end{bmatrix}$$

and

$$x(\vec{t}) = \Pi\left[\frac{1}{\sqrt{2}}\left(t_1 + \frac{t_2}{2}\right) - \frac{1}{2}\right] \Pi\left[\frac{1}{\sqrt{2}}\left(t_1 - \frac{t_2}{2}\right) + \frac{1}{2}\right]$$

6.1.1.2.3   Fourier Transformation

To determine the effects of rotation and scale on a function's Fourier transform, we evaluate

$$x(\mathbf{A}\vec{t}) \quad \leftrightarrow \quad \int_{\vec{t}} x(\mathbf{A}\vec{t}) \exp(-j2\pi\vec{u}^T\vec{t})d\vec{t} \tag{6.11}$$

$$= \quad \frac{1}{|\det \mathbf{A}|} \int_{\vec{\tau}} x(\vec{\tau}) \exp(-j2\pi(\mathbf{A}^{-T}\vec{u})^T\vec{\tau})d\vec{\tau} \tag{6.12}$$

$$= \quad \frac{X(\mathbf{A}^{-T}\vec{u})}{|\det \mathbf{A}|} \tag{6.13}$$

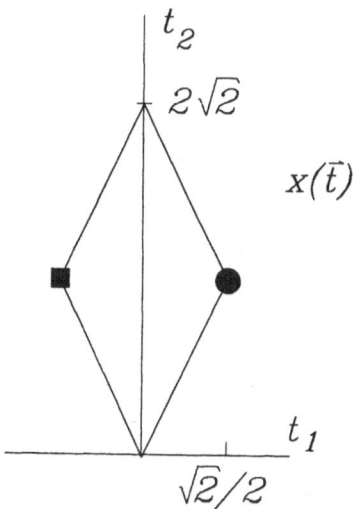

Figure 6.5: Rotating the unit square in Figure 6.1 by 45° and scaling by a factor of 2 in the t_2 direction.

where we have made the substitution $\vec{\tau} = \mathbf{A}\vec{t}$ and $|\det \mathbf{A}|$ is the transformation Jacobian. We assume $\mathbf{A}$ is not singular. The superscript $-T$ denotes inverse transposition.

Example : For the rotation matrix in (6.10), $R^{-1} = R^T$. Therefore, rotating a function rotates its transform. This is also true in higher dimensions.

Example : For the scaling matrix in (6.8), we conclude that

$$x\left(\frac{t_1}{M_1}, \frac{t_2}{M_2}\right) \leftrightarrow M_1 M_2 X(M_1 u_1, M_2 u_2)$$

More generally, if $\mathbf{S}$ is the diagonal matrix in (6.9), we have

$$x(\mathbf{A}\vec{t}) \leftrightarrow M_1 M_2 \cdots M_N X(M_1 u_1, M_2 u_2, \cdots, M_N u_N)$$

Example : Since

$$y(\vec{t}) = \Pi\left(t_1 - \frac{1}{2}\right) \Pi\left(t_2 - \frac{1}{2}\right) \leftrightarrow \operatorname{sinc}(u_1)\operatorname{sinc}(u_2)\exp[-j\pi(u_1+u_2)]$$

we conclude from (6.13) that the Fourier transform of the parallelogram at the top of Figure 6.2 is

$$x(\vec{t}) = y(\mathbf{A}\vec{t}) \leftrightarrow 4\mathrm{sinc}(3u_1 - u_2)\mathrm{sinc}(u_1 + u_2)\exp(-j4\pi u_1)$$

The results of the above examples can be combined to visualize the effects of rotation and scaling on a function's transform. Rotation follows in the transform domain whereas a compression in one dimension is countered by an expansion along the corresponding frequency dimension.

6.1.1.3   Polar Representation

Choosing other coordinates can alter the mechanics of performing a Fourier transform. Consider, for example, expressing a two dimensional function in polar coordinates as $x(r, \theta)$. The function is periodic in θ with a period of 2π and can therefore be represented as a Fourier series

$$x(r,\theta) = \sum_{n=-\infty}^{\infty} c_n(r)e^{jn\theta} \qquad (6.14)$$

where

$$c_n(r) = \frac{1}{2\pi} \int_0^{2\pi} x(r, \theta)e^{-jn\theta}d\theta \qquad (6.15)$$

are referred to as *circular harmonics*.

The two dimensional polar equivalent of (6.1) can be obtained by the variable substitutions

$$t_1 = r\cos(\theta); \ t_2 = r\sin(\theta)$$

$$dt_1 dt_2 = rdrd\theta$$

and the polar frequency coordinates, (ρ, ϕ), related by

$$u_1 = \rho\cos(\phi); \ u_2 = \rho\sin(\phi)$$

The result is

$$X(\rho, \phi) = \int_{\theta=0}^{2\pi} \int_{r=0}^{\infty} x(r, \theta)\exp[j2\pi r\rho\cos(\theta - \phi)]rdrd\theta \qquad (6.16)$$

where we have used a trigonometric identity. Substituting the circular harmonic expansion in (6.14) and using the identity

$$J_n(z) = \frac{1}{2\pi} \int_{-\pi}^{\pi} e^{jn\phi} e^{jz\sin(\phi)} d\phi$$

gives, after some manipulation,

$$X(\rho, \phi) = \sum_{n=-\infty}^{\infty} j^n e^{jn\phi} \mathcal{H}_n[c_n(r)] \qquad (6.17)$$

where the nth order Hankel transform is defined by

$$\mathcal{H}_n[f(r)] = 2\pi \int_0^{\infty} r f(r) J_n(2\pi r\rho) dr \qquad (6.18)$$

Note, in particular, that (6.17) is a polar Fourier series of the spectrum of $X(\rho, \phi)$ with circular harmonics $(-j)^n \mathcal{H}_n[c_n(r)]$.

6.1.1.3.1   The Hankel Transform

A zeroth order Hankel transform is referred to as a *Fourier-Bessel* or simply Hankel transform. From (6.18)

$$\mathcal{H}f(r) = \mathcal{H}_0 f(r) = 2\pi \int_0^{\infty} r f(r) J_0(2\pi r\rho) dr \qquad (6.19)$$

Consider the case where a two dimensional function is circularly symmetric. That is $x(r, \theta) = f(r)$. Then, from (6.14), $c_o(r) = f(r)$ and all other circular harmonics are zero. The series in (6.17) thus reduces to $\mathcal{H}f(r)$. Note that, in such a case, the transform is not a function of its angular variable and, with $F(\rho) = X(\rho, \phi)$, we can write

$$F(\rho) = \mathcal{H}f(r)$$

In other words, if a two dimensional function is circularly symmetric, then so is its two dimensional Fourier transform. Furthermore, the two dimensional transform is equivalent to a (one dimensional) Hankel transform. Generalizations of Hankel transformations in dimensions greater than two are discussed by Bracewell.

6.1.1.3.2   Example

Consider the unit radius circle

$$f(r) = \Pi\left(\frac{r}{2}\right) \tag{6.20}$$

Substituting into (6.19) gives

$$F(\rho) = 2\pi \int_0^1 r J_0(2\pi r\rho)dr$$

Since

$$\frac{d}{d\alpha}\alpha J_1(\alpha) = \alpha J_0(\alpha)$$

we conclude that

$$F(\rho) = 2\mathrm{jinc}(\rho) \tag{6.21}$$

6.1.2   Fourier Series

Multidimensional periodic functions can be expanded in a multidimensional generalization of the Fourier series. First, we must establish some formality for characterizing multidimensional periodicity.

6.1.2.1   Multidimensional Periodicity

An N dimensional function, $S(\vec{u})$, is said to be periodic if there exists a set of N non-colinear vectors, $\{\vec{p}_n | 1 \leq n \leq N\}$, such that, for all u,

$$S(\vec{u}) = S(\vec{u} - \vec{p}_1) = S(\vec{u} - \vec{p}_2) = \cdots = S(\vec{u} - \vec{p}_N) \tag{6.22}$$

The matrix

$$\mathbf{P} = [\vec{p}_1 | \vec{p}_2 | \cdots | \vec{p}_N]$$

is referred to as the *periodicity matrix*. Requiring that the $\vec{p}_n$'s not be colinear is equivalent to requiring that $\mathbf{P}$ not be singular ($\det \mathbf{P} \neq 0$).

Example : A periodic $N = 2$ signal is shown in Figure 6.6 with

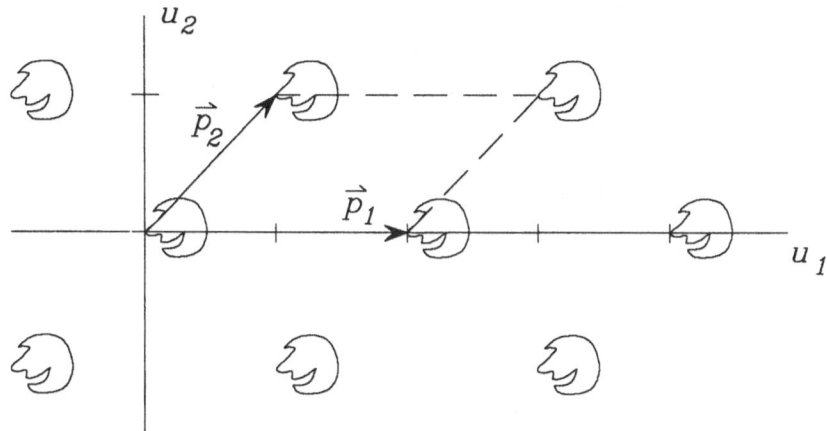

Figure 6.6: A two dimensional periodic function is one inside the replicated closed curve and zero outside.

periodicity matrix

$$\mathbf{P} = \begin{bmatrix} 1 & 2 \\ 1 & 0 \end{bmatrix} \tag{6.23}$$

The concept of a *period* is not as straightforward as in one dimension since we now must deal with geometrical forms rather than with simple intervals. A *period cell* is any region that, when replicated in accordance with the periodicity matrix, will fill the entire space without gaps. Two of many possible cells for the replication geometry in Figure 6.6 are shown in Figure 6.7. Although there are many possible choices, every cell must have the same area.[1]

A periodicity cell can always be formed from the multidimensional equivalent of the parallelogram defined by the periodicity vectors. The parallelogram formed by the periodicity vector and the dashed lines in Figure 6.6, for example, clearly form a cell

---

[1]In this chapter, area means, in general, multidimensional area, *e.g.*, for $N = 3$, volume.

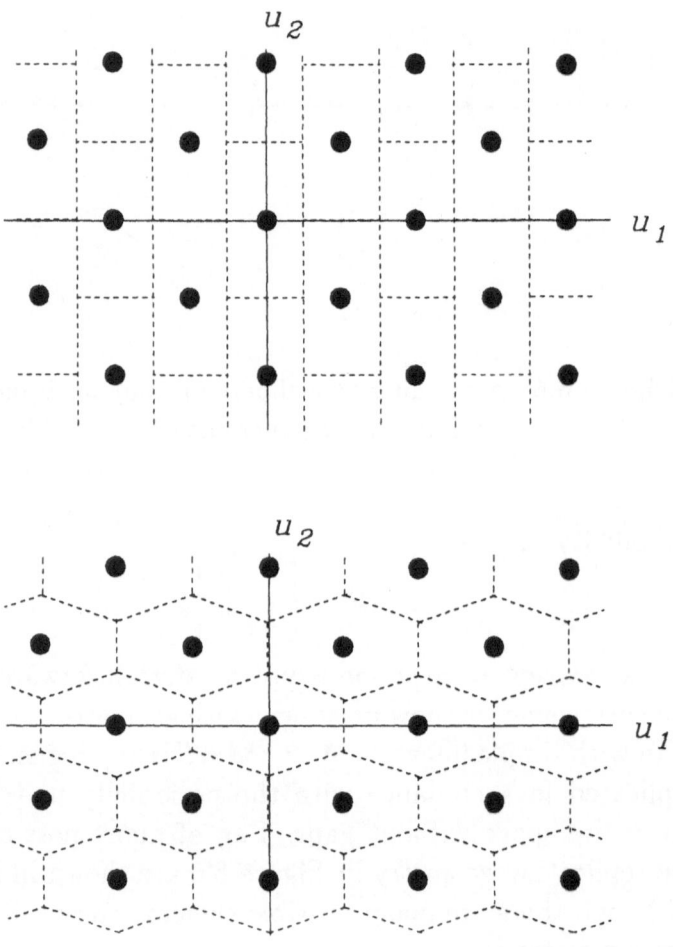

Figure 6.7: For a given replication geometry, there may exist many ways to partition the plane with periodicity cells. Shown here are two such partitionings for the replication geometry shown in Figure 6.6. Solid dots, shown at the center of each cell, are identically positioned in both partitions.

for the replication geometry shown.

Note that there exists one replication point per cell. The number of periods per unit area, T, is thus equal to the area of a cell which can be calculated from the parallelogram configuration as

$$T = |\det \mathbf{P}|$$

The periodicity in Figure 6.6, for example, has 2 cells per unit area.

6.1.2.2   The Fourier Series Expansion

In this section, we derive the Fourier series for a periodic function, $S(\vec{u})$, with given periodicity matrix $\mathbf{P}$. First we establish as a Fourier transform pair

$$\sum_{\vec{m}} \delta(\vec{t} - \mathbf{Q}\vec{m}) \leftrightarrow |\det \mathbf{P}| \sum_{\vec{m}} \delta(\vec{u} - \mathbf{P}\vec{m}) \qquad (6.24)$$

where

$$\mathbf{Q}^T = \mathbf{P}^{-1} \qquad (6.25)$$

and the multidimensional Dirac delta is

$$\delta(\vec{t}) = \delta(t_1)\delta(t_2) \cdots \delta(t_N)$$

Proof :

If we define

$$\lambda(\vec{t}) = \sum_{\vec{m}} \delta(\vec{t} - \mathbf{Q}\vec{m})$$

then

$$\lambda(\mathbf{Q}\vec{t}) = \frac{\sum_{\vec{m}} \delta(\vec{t} - \vec{m})}{|\det \mathbf{Q}|} = |\det \mathbf{P}| \prod_{n=1}^{N} \text{comb}(t_n)$$

where we have used the identities

$$|\det \mathbf{P}| = \frac{1}{|\det \mathbf{Q}|}$$

and

$$\delta(\mathbf{A}\vec{t}) = \frac{\delta(\vec{t})}{|\det \mathbf{A}|} \qquad (6.26)$$

Using the last entry in Table 2.2 and the separability theorem, we conclude that

$$\lambda(\mathbf{Q}\vec{t}) \leftrightarrow |\det \mathbf{P}| \prod_{n=1}^{N} \text{comb}(u_n) = |\det \mathbf{P}| \sum_{\vec{m}} \delta(\vec{u} - \vec{m})$$

Using the rotation and scale theorem, we find that

$$\lambda(\vec{t}) \leftrightarrow \sum_{\vec{m}} \delta(\mathbf{Q}^T \vec{u} - \vec{m})$$

from which (6.24) immediately follows.

$$* * * * * *$$

Using (6.24), we can now show that a periodic function, $S(\vec{u})$ with periodicity matrix $\mathbf{P}$ can be expressed via the Fourier series expansion

$$S(\vec{u}) = \sum_{\vec{m}} c[\vec{m}] \exp(-j2\pi \vec{u}^T \mathbf{Q}\vec{m}) \qquad (6.27)$$

where the Fourier coefficients are

$$c[\vec{m}] = |\det \mathbf{Q}| \int_{\vec{u} \in \mathcal{C}} S(\vec{u}) \exp(j2\pi \vec{u}^T \mathbf{Q}\vec{m}) d\vec{u} \qquad (6.28)$$

and $\mathcal{C}$ is any periodicity cell.

Proof :
Let $|\det \mathbf{P}| X(\vec{u}) = S(\vec{u})$ over any cell region and be zero elsewhere. This function is equivalent to a period of $S(\vec{u})$. We can therefore replicate it to form $S(\vec{u})$:

$$S(\vec{u}) = |\det \mathbf{P}| X(\vec{u}) * \sum_{\vec{m}} \delta(\vec{u} - \mathbf{P}\vec{m}) \qquad (6.29)$$

Using the modulation theorem in Table 6.1, and Equation (6.24), we conclude

$$s(\vec{t}) = x(\vec{t}) \sum_{\vec{m}} \delta(\vec{t} - \mathbf{Q}\vec{m}) = \sum_{\vec{m}} x(\mathbf{Q}\vec{m}) \delta(\vec{t} - \mathbf{Q}\vec{m})$$

where

$$x(\vec{t}) \leftrightarrow X(\vec{u}) \qquad (6.30)$$

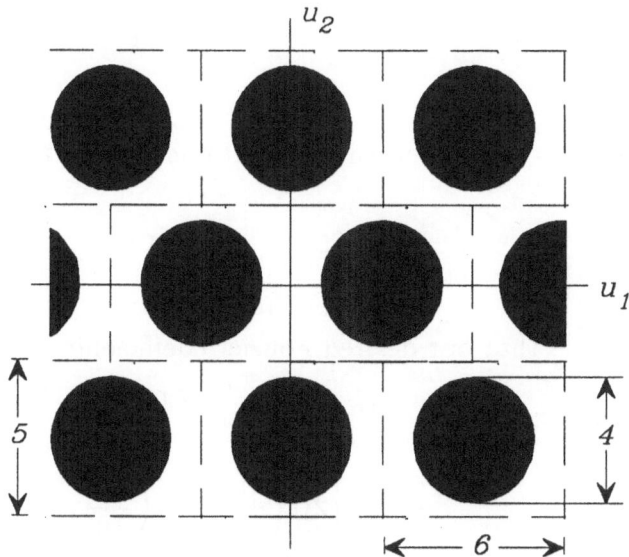

Figure 6.8: A periodic function for which we find a Fourier series. The function is zero inside the circles and one outside.

and

$$s(\vec{t}) \leftrightarrow S(\vec{u})$$

Transforming gives

$$S(\vec{u}) = \sum_{\vec{m}} x(\mathbf{Q}\vec{m}) \exp(-j2\pi \vec{u}^T \mathbf{Q}\vec{m}) \qquad (6.31)$$

and, after noting that $c[\vec{m}] = x(\mathbf{Q}\vec{m})$ our proof is complete.

Example : Consider the periodic function, $S(\vec{u})$, shown in Figure 6.8. The function is zero in the shaded circles and one outside. A periodicity matrix for this function is

$$\mathbf{P} = \left[\begin{array}{cc} 6 & 3 \\ 0 & 5 \end{array} \right]$$

We choose the rectangular periodicity cells shown with the dashed lines. To find the Fourier coefficients, we choose the cell centered at $(3,0)$. Correspondingly,

$$|\det \mathbf{P}|X(\vec{u}) = \Pi\left(\frac{u_1 - 3}{6}\right)\Pi\left(\frac{u_2}{5}\right) - \Pi\left[\frac{\sqrt{(u_1 - 3)^2 + u_2^2}}{4}\right]$$

Therefore

$$|\det \mathbf{P}|x(\vec{t}) = [30\text{sinc}(6t_1)\text{sinc}(5t_2) - 8\text{jinc}(2\sqrt{t_1^2 + t_2^2})]\exp(-j6\pi t_1)$$

and, since

$$\mathbf{Q}\vec{m} = \left[\begin{array}{c} \frac{m_1}{6} \\ -\frac{m_1}{10} + \frac{m_2}{5} \end{array}\right]$$

we conclude that our desired Fourier coefficients are

$$\begin{aligned} c[m_1, m_2] = & (-1)^{m_1}\left\{\delta[m_1]\text{sinc}\left(\frac{m_1}{2} - m_2\right)\right. \\ & \left. - \frac{4}{15}\text{jinc}\left[2\sqrt{\left(\frac{m_1}{6}\right)^2 + \left(\frac{m_1}{10} - \frac{m_2}{5}\right)^2}\right]\right\} \end{aligned}$$

The resulting Fourier series using these coefficients is

$$S(u_1, u_2) = \sum_{m_1=-\infty}^{\infty}\sum_{m_2=-\infty}^{\infty} c[m_1, m_2]\exp\left[-j2\pi\left(\frac{m_1 u_1}{6} - \frac{m_1 u_2}{10} + \frac{m_2 u_2}{5}\right)\right]$$

6.2   The Multidimensional Sampling Theorem

We have established in one dimension that the Shannon sampling theorem is the Fourier dual of a Fourier series. The same is true in higher dimensions. Indeed, we can glean the multidimensional sampling theorem from our development of the Fourier series by interpreting

- $x(\vec{t})$ as a multidimensional bandlimited function

- $X(\vec{u})$ as its spectrum. The function $x(\vec{t})$ is defined to be bandlimited if $X(\vec{u}) = 0$ outside of an N-dimensional sphere of finite radius.

- $S(\vec{u})$ as the replication of $|\det \mathbf{P}|X(\vec{u})$ as a result of sampling.

Clearly, we must choose our periodicity matrix, $\mathbf{P}$, so that the spectral replications do not overlap and therefore alias. We then can always choose a periodicity cell, $\mathcal{C}$, so that

$$S(\vec{u}) = |\det \mathbf{P}|X(\vec{u}); \ \vec{u} \in \mathcal{C} \tag{6.32}$$

From (6.28),
$$c[\vec{m}] = x(\mathbf{Q}\vec{m})$$

and the Fourier series in (6.27) becomes

$$S(\vec{u}) = \sum_{\vec{m}} x(\mathbf{Q}\vec{m})\exp(-j2\pi\vec{u}^T\mathbf{Q}\vec{m})$$

Using (6.32) and inverse transforming over a cell gives

$$x(\vec{t}) = \sum_{\vec{m}} x(\mathbf{Q}\vec{m})f_C(\vec{t} - \mathbf{Q}\vec{m}) \tag{6.33}$$

where

$$f_C(\vec{t}) = |\det \mathbf{Q}| \int_{\vec{u} \in C} \exp(j2\pi\vec{u}^T\vec{t})d\vec{u} \tag{6.34}$$

Equations (6.33) and (6.34) summarize the multidimensional sampling theorem.

The $\mathbf{Q}$ matrix is called the *sampling matrix* and dictates the geometry of the uniform sampling. We decompose $\mathbf{Q}$ into vector columns:

$$\mathbf{Q} = [\vec{q}_1|\vec{q}_2|\cdots|\vec{q}_N] \tag{6.35}$$

As with the periodicity matrix, each component vector is a basis vector for sampling. An example is shown in Figure 6.9 for $N = 2$ and

$$\mathbf{Q} = \begin{bmatrix} -1 & 2 \\ 3 & -2 \end{bmatrix} \tag{6.36}$$

It follows that

$$\mathbf{P} = \begin{bmatrix} \frac{1}{2} & \frac{3}{4} \\ \frac{1}{2} & \frac{1}{4} \end{bmatrix}$$

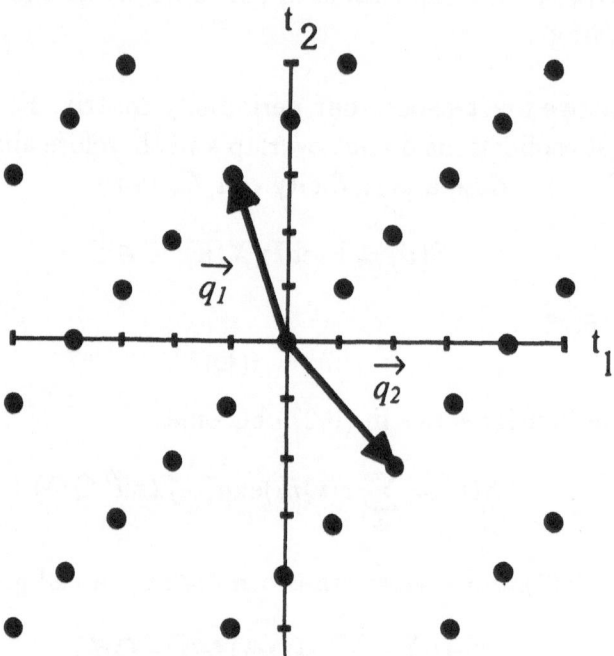

Figure 6.9: Sampling geometry corresponding to the sampling matrix in (6.36).

An example of the spectral replication using this matrix is shown in Figure 6.10.

Consider the parallelogram defined by the column vectors of $\mathbf{Q}$. Visualize replication to fill the entire plane. Clearly, there is a one to one correspondence of a sample to a parallelogram. Since the area of the parallelogram is $|\det \mathbf{Q}|$, we conclude that the *sampling density* corresponding to $\mathbf{Q}$ is

$$\text{SD} = \frac{1}{|\det \mathbf{Q}|} = |\det \mathbf{P}| \text{ samples/unit area}$$

The sampling geometry in Figure 6.9, for example, has a density of one fourth of a sample per unit area.

We call the region $\mathcal{A}$, over which a spectrum $X(\vec{u})$ is not identically zero, the spectral *support*. In order for $x(\vec{t})$ to follow our definition of bandlimitedness, $\mathcal{A}$ must be totally contained within an N dimensional sphere of finite radius.

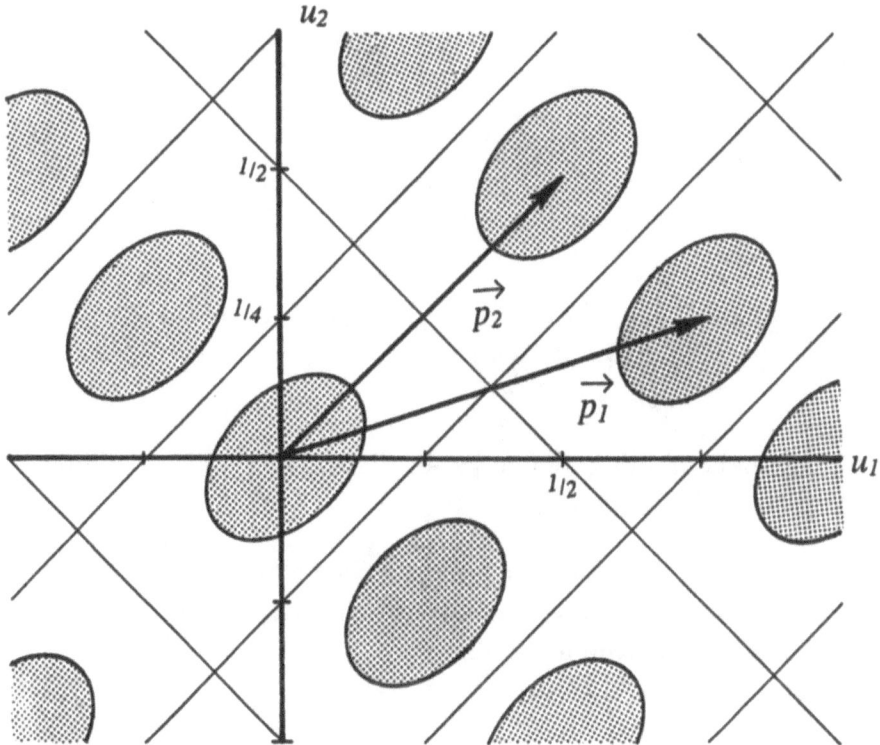

Figure 6.10: Spectrum replication from the sampling geometry of Figure 6.9.

6.2.1   The Nyquist Density

The lowest rate at which a temporal bandlimited signal can be sampled without aliasing is the *Nyquist rate*. This measure can be generalized to higher dimensions. For a given spectral support, $\mathcal{A}$, the *Nyquist density* is that density resulting from maximally packed unaliased replication of the signal's spectrum.

Some illustrative examples are in order.

Example a : If the support of an $N = 2$ dimensional spectrum has the rectangular shape shown on the left of Figure 6.11, then

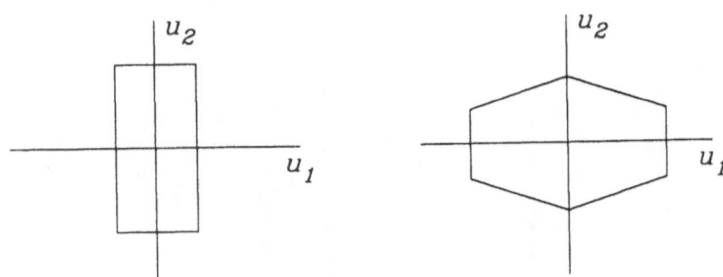

Figure 6.11: Two spectral supports. The value outside a support region is identically zero.

an optimal sampling geometry results in the replication shown at the top of Figure 6.7.

Example b : For the hexagonal support shown at the right of Figure 6.11, the Nyquist density is achieved by the replication shown at the bottom of Figure 6.7.

Notes:

- In Example **a**, there is more than one sampling geometry that can achieve the Nyquist density. Placing the rectangles in a chess board grid is another possibility. There is only one way, on the other hand, to replicate the hexagon in Example **b** to efficiently fill the plane.

- For both Example **a** and **b**, the support of the spectrum, A, is of the shape of a cell, C. In such instances, the Nyquist density is equal to the area of the spectral support.

Example c : Any imaging system that uses lenses with circular pupils will generate images the spectra of which have a circular region of support if the monochromatic illumination is either coherent or incoherent [Goodman] [Gaskill]. Consider, then, the case where the support of an $N = 2$ dimensional spectrum is a

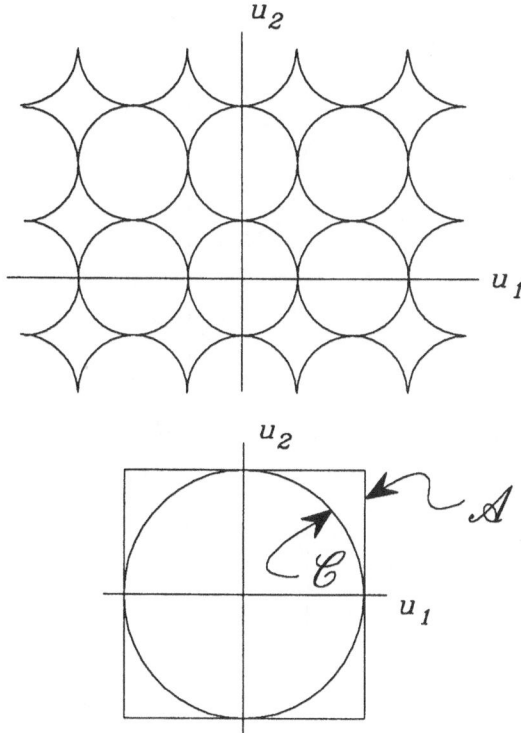

Figure 6.12: (Top) Minimum density rectangular sampling of images with spectra of circular support yields circles packed as shown. (Bottom) A single cell with inscribed circular spectrum support.

circle of radius W. If we limit ourselves to rectangular sampling (i.e., restrict $\mathbf{Q}$ to be a diagonal matrix), then the closest we can pack circles is illustrated at the top of Figure 6.12. The periodicity and sampling matrices are, respectively,

$$\mathbf{P} = \begin{bmatrix} 2W & 0 \\ 0 & 2W \end{bmatrix}$$

$$\mathbf{Q} = \begin{bmatrix} T & 0 \\ 0 & T \end{bmatrix}$$

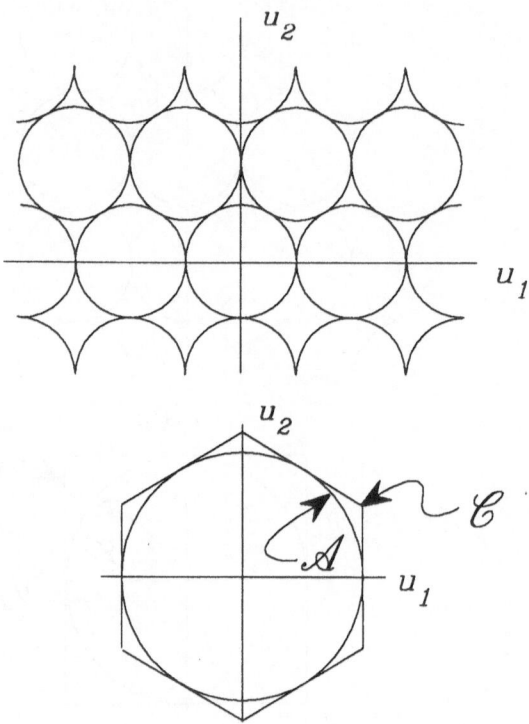

Figure 6.13: (Top) Densely packed circles correspond to Nyquist sampling of images with spectra of circular support. Note the hexagonal structure. (Bottom) A single hexagonal cell with inscribed circular spectrum support.

where

$$T = \frac{1}{2W}$$

The cell, $\mathcal{C}$, corresponding to rectangular sampling is the square shown at the bottom of Figure 6.12.

Example d : Maximally dense packing of circles is shown in Figure 6.13. The corresponding density matrix is

$$\mathbf{P} = \begin{bmatrix} W & -W \\ \sqrt{3}W & \sqrt{3}W \end{bmatrix} \qquad (6.37)$$

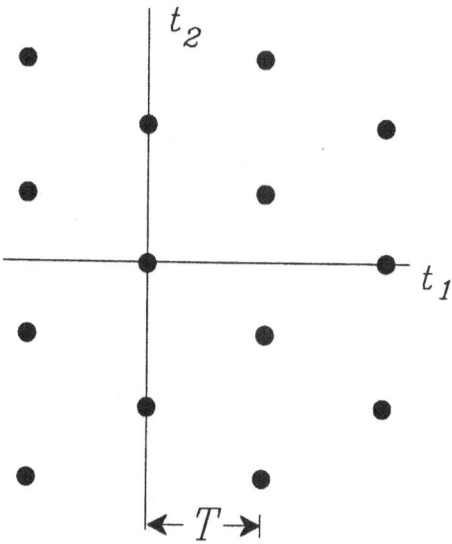

Figure 6.14: Hexagonal sampling geometry required to pack circles densely as shown in Figure 6.13.

The sampling matrix follows as

$$\mathbf{Q} = \begin{bmatrix} T & -T \\ \frac{T}{\sqrt{3}} & \frac{T}{\sqrt{3}} \end{bmatrix}$$

As shown in Figure 6.14, this sampling geometry is hexagonal in nature. The cell resulting from such sampling is the hexagon shown at the bottom of Figure 6.13.

* * * * * * *

What is the savings in sampling density of the hexagonal geometry in Example **d** over the rectangular geometry in Example **c**? For hexagonal sampling, the (Nyquist) sampling density is equal to the area of the hexagon shown at the bottom of Figure 6.13 which in turn, is equal to the determinant magnitude of (6.37):

$$SD_{nyq} = 2\sqrt{3}W^2 \tag{6.38}$$

The corresponding density for rectangular sampling is clearly

$$SD_{rect} = 4W^2$$

The ratio of these densities is

$$r_2 = \frac{SD_{nyq}}{SD_{rect}} = 0.866$$

where the subscript denotes $N = 2$ dimensions. Thus, use of hexagonal sampling reduces the sampling density by 13.4% over rectangular sampling.

This result has been extended to higher dimensions for an N dimensional hypersphere of radius W [Peterson and Middleton] [Dudgeon and Mersereau]. The minimum rectangular sampling density is $SD_{rect} = (2W)^N$. The Nyquist density, SD_{nyq}, can be evaluated from the geometry of maximally packed spheres. The ratio

$$r_N = \frac{SD_{nyq}}{SD_{rect}}$$

is plotted in Figure 6.15 up to eight dimensions. The difference in sampling densities clearly becomes more significant in higher dimensions. The numerical values from which this plot was made are in Table 6.2. We will need them later.

6.2.2   Generalized Interpolation Functions

In this section, we extend some of the results of Section 4.1 to higher dimensions.

6.2.2.1   Tightening the Integration Region

A single cell for the replication in Figure 6.10 is shown in Figure 6.16. Clearly, the support of a signal's spectrum, $\mathcal{A}$, will be subsumed in a periodicity cell, $\mathcal{C}$. Since the spectrum is identically zero outside of the support region, $\mathcal{A}$, we can rewrite the interpolation formulae in (6.33) and (6.34) as

$$x(\vec{t}) = \sum_{\vec{m}} x(Q\vec{m})f_B(\vec{t} - Q\vec{m}) \tag{6.39}$$

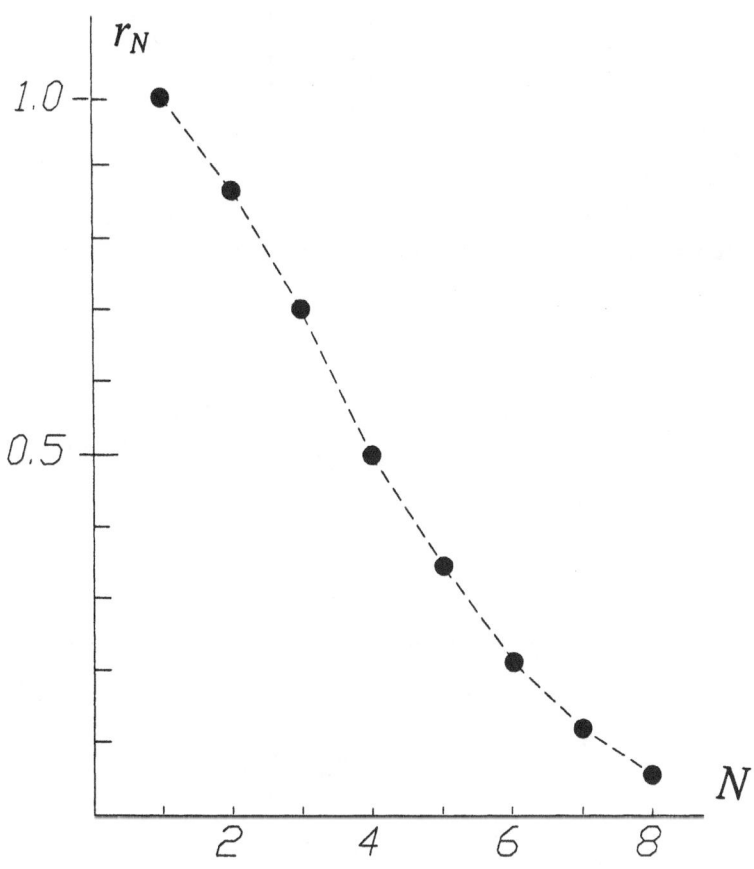

Figure 6.15: The ratio of the Nyquist density to the minimum rectangular sampling density when the support of the signal's spectrum is an N dimensional hypercube.

N	r_N
1	1.000
2	0.866
3	0.705
4	0.499
5	0.353
6	0.217
7	0.125
8	0.062

Table 6.2: The numbers from which Figure 6.15 was generated.

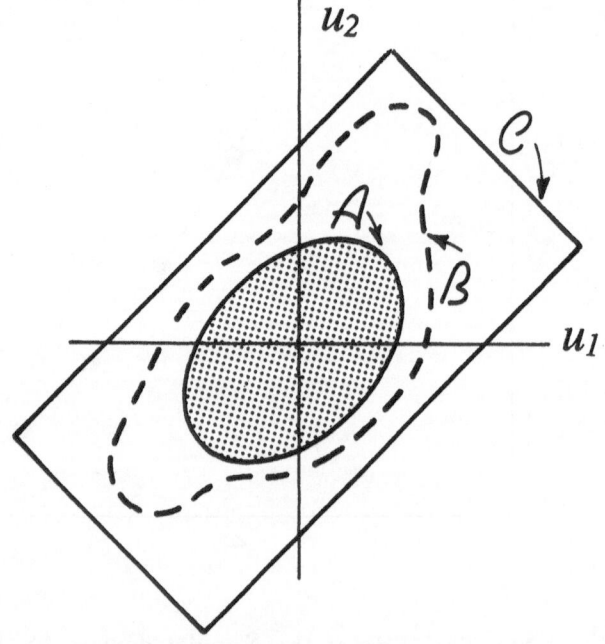

Figure 6.16: One cell of Figure 6.10. The region of integration, $\mathcal{B}$, must co ntain the spectral support region, $\mathcal{A}$, and must not infringe onto adjacent spectra. $\mathcal{C}$ is a cell region. The areas of the regions $\mathcal{A}, \mathcal{B}$ and $\mathcal{C}$ are A, B and C respectively.

and

$$f_B(\vec{t}) = |\det \mathbf{Q}| \int_{\vec{u} \in B} \exp(j2\pi \vec{u}^T \vec{t}) d\vec{u} \qquad (6.40)$$

where B is any region within a periodicity cell that contains the support, A. Indeed, the region B could be the support region, A, or, as in (6.34), a periodicity cell, C. Note that, in one dimension, $A = B = C$ at the Nyquist rate. In higher dimensions, such an equality is not assured at the Nyquist density.

6.2.2.2   Allowing Slower Roll Off

In one dimension, we were able to allow the spectrum of the interpolation function to be any desired function over those intervals where the signal's spectrum was identically zero. This allowed use of interpolation functions whose spectra had a slower roll off (see Figure 4.2). An analogous freedom occurs in higher dimensions. We can write, for example

$$x(\vec{t}) = \sum_{\vec{m}} x(\mathbf{Q}\vec{m}) f_B^+(\vec{t} - \mathbf{Q}\vec{m}) \qquad (6.41)$$

where

$$f_B^+(\vec{t}) = f_B(\vec{t}) + \int_{C \cap \bar{B}} \Re(\vec{u}) \exp(j2\pi \vec{u}^T \vec{t}) d\vec{u} \qquad (6.42)$$

and $\cap$ denotes intersection. The function $\Re(\vec{u})$ can be any convenient function. Integration is over all points in C not contained in B.

6.3   Restoring Lost Samples

In this section, we will show that an arbitrarily large but finite number of lost samples can be regained from those remaining for certain band-limited signals even when sampling is performed at the Nyquist density [Marks].

6.3.1   Restoration Formulae

Let $\mathcal{M}$ denote a set of M integer vectors corresponding to the M lost-sample locations in an N- dimensional bandlimited sig-

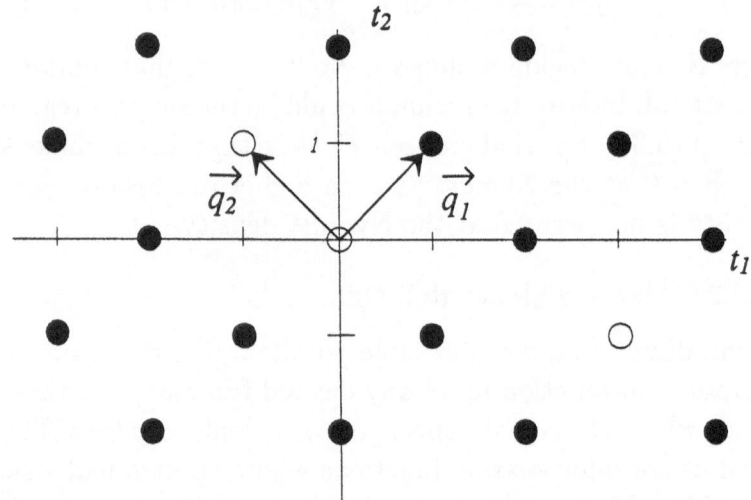

Figure 6.17: Illustration of the three lost samples in the set $\mathcal{M} = \{\vec{m}_1, \vec{m}_2, \vec{m}_3\}$. The location of the three lost samples is $\mathbf{Q}\vec{m}_k$; $k = 1, 2, 3$.

nal sampled in accordance with a sampling matrix, $\mathbf{Q}$. As an example, consider Figure 6.17 where the sampling matrix is

$$\mathbf{Q} = \begin{bmatrix} 1 & -1 \\ 1 & 1 \end{bmatrix}$$

A total of $M = 3$ lost samples are shown by hollow dots. It follows that

$$\mathcal{M} = \left\{ \begin{bmatrix} 0 \\ 0 \end{bmatrix}, \begin{bmatrix} 0 \\ 1 \end{bmatrix}, \begin{bmatrix} 1 \\ -2 \end{bmatrix} \right\}$$

Theorem : If $x(\vec{t})$ is a bandlimited signal and $\mathbf{Q}$ is chosen to ensure that there is no aliasing, then the missing samples can be regained from solution of the M equations:

$$\sum_{\vec{n} \notin \mathcal{M}} x(\mathbf{Q}\vec{n})\{\delta[\vec{k} - \vec{n}] - f_B[\mathbf{Q}(\vec{k} - \vec{n})]\} = \sum_{\vec{n} \in \mathcal{M}} x(\mathbf{Q}\vec{n})f_B[\mathbf{Q}(\vec{k} - \vec{n})]$$

$$\vec{k} \in \mathcal{M} \qquad\qquad (6.43)$$

assuming that the solution is not singular. [The Kronecker delta function, $\delta[\vec{n}]$, is unity when $\vec{n} = \vec{0}$ and is zero otherwise.] The left-hand side of (6.43) contains the unknown samples. The right-hand side can be found from the known data.

Proof: We can write (6.39) as

$$x(\vec{t}) = \left[\sum_{\vec{n} \notin \mathcal{M}} + \sum_{\vec{n} \in \mathcal{M}}\right] x(\mathbf{Q}\vec{n}) f_B(\vec{t} - \mathbf{Q}\vec{m})$$

This expression can be evaluated at M points, and we can solve for the sample set $\{x(\mathbf{Q}\vec{n}|\vec{n} \in \mathcal{M}\}$. Let these M points be the $\vec{t} = \mathbf{Q}\vec{k}$, where $\vec{k} \in \mathcal{M}$:

$$x(\mathbf{Q}\vec{k}) = [\sum_{\vec{n} \notin \mathcal{M}} + \sum_{\vec{n} \in \mathcal{M}}] x(\mathbf{Q}\vec{n}) f_B\{\mathbf{Q}(\vec{k} - \vec{n})\} ; \quad \vec{k} \in \mathcal{M}$$

Rearranging gives (6.43).

Corollary 1 : For a single lost sample at the origin, if $f_B(\vec{0}) \neq 1$, then

$$x(\vec{0}) = [1 - f_B(\vec{0})]^{-1} \sum_{\vec{n} \neq \vec{0}} x(\mathbf{Q}\vec{n}) f_B(-\mathbf{Q}\vec{n}) \qquad (6.44)$$

This follows from (6.43) for $M = 1$ and $\mathcal{M}$ containing only the origin. Note that, by using (6.39), the signal's interpolation can be written directly void of the sample at the origin:

$$x(\vec{t}) = \sum_{\vec{n} \neq \vec{0}} x(\mathbf{Q}\vec{n})[f_B(\vec{t} - \mathbf{Q}\vec{n}) + \{1 - f_B(\vec{0})\}^{-1} f_B(-\mathbf{Q}\vec{n}) f_B(\vec{t})]$$

Corollary 2 : A sufficient condition for (6.43) to be singular is when the integration region, $\mathcal{B}$, is equal to a cell region, $\mathcal{C}$.

Proof : On a cell, the functions $\{\exp(j2\pi\vec{u}^T\mathbf{Q}\vec{n})\}$ form an orthogonal basis set. From (6.40) with $\mathcal{B} = \mathcal{C}$ we have

$$f_B(\mathbf{Q}\vec{n}) = |\det \mathbf{Q}| \int_{\mathcal{C}} \exp(j2\pi\vec{u}^T\mathbf{Q}\vec{n}) d\vec{u} = \delta[\vec{n}]$$

The left-hand side of (6.43) is thus zero and the resulting set of equations singular.

6.3.2 Noise Sensitivity

Our purpose here is to investigate the restoration algorithm's performance when inaccurate data are used. In general, the algorithm becomes more unstable when (1) M increases and/or (2) the area corresponding to $\mathcal{B}$ increases with respect to that of $\mathcal{C}$. Indeed, restoration is no longer possible when $\mathcal{B} = \mathcal{C}$.

The restoration algorithm in (6.43) is linear. Let $\xi(\vec{t})$ denote a zero mean stochastic process. If $x(\vec{t})$ is uncorrelated with $\xi(\vec{t})$, then the use of $\{x(\mathbf{Q}\vec{n}) + \xi(\mathbf{Q}\vec{n})|\vec{n} \notin \mathcal{M}\}$ in (6.43) instead of $\{x(\mathbf{Q}\vec{n})|\vec{n} \notin \mathcal{M}\}$ will result in $\{x(\mathbf{Q}\vec{n}) + \eta(\mathbf{Q}\vec{n})|\vec{n} \in \mathcal{M}\}$, where $\{\eta(\mathbf{Q}\vec{n})|\vec{n} \in \mathcal{M}\}$ is the response to $\{\xi(\mathbf{Q}\vec{n})|\vec{n} \notin \mathcal{M}\}$ alone:

$$\sum_{\vec{n} \notin \mathcal{M}} \eta(\mathbf{Q}\vec{n})\{\delta[\vec{k} - \vec{n}] - f_{\mathcal{B}}\{\mathbf{Q}(\vec{k} - \vec{n})\} = \sum_{\vec{n} \in \mathcal{M}} \xi(\mathbf{Q}\vec{n})f_{\mathcal{B}}\{\mathbf{Q}(\vec{k} - \vec{n})\}$$
$$\vec{k} \in \mathcal{M} \tag{6.45}$$

The restoration noise, η, depends linearly on the data noise, ξ. Thus the cross correlation between these two processes and the autocorrelation of η can be determined from a given data noise autocorrelation.

Our treatment will be limited to the case when a single sample is lost and the data noise is white, i.e.,

$$E[\xi(\mathbf{Q}\vec{n})\xi^*(\mathbf{Q}\vec{m})] = \overline{\xi^2}\delta[\vec{n} - \vec{m}] \tag{6.46}$$

where $\overline{\xi^2}$ is the data noise level (variance) and E denotes expectation. With no loss in generality, we place the lost sample at the origin, and (6.45) becomes

$$\eta(\vec{0}) = [1 - f_{\mathcal{B}}(\vec{0})]^{-1} \sum_{\vec{n} \neq \vec{0}} \xi(\mathbf{Q}\vec{n})f_{\mathcal{B}}(-\mathbf{Q}\vec{n})$$

Taking the square of the magnitude, expectating, and using (6.46) gives

$$\frac{\overline{\eta^2(\vec{0})}}{\overline{\xi^2}} = [1 - f_{\mathcal{B}}(\vec{0})]^{-2} \sum_{\vec{n} \neq \vec{0}} |f_{\mathcal{B}}(-\mathbf{Q}\vec{n})|^2 \tag{6.47}$$

where the restoration noise level is

$$\overline{\eta^2(\vec{0})} = E[|\eta^2(\vec{0})|]$$

The sum in (6.47) can be evaluated through (6.44) with $x(\vec{t}) = f^*(-\vec{t})$ [$= f(\vec{t})$ since $F_B(\vec{u})$ is real]. The result is

$$\frac{\overline{\eta^2(\vec{0})}}{\overline{\xi^2}} = \frac{f_B(\vec{0})}{1 - f_B(\vec{0})} \tag{6.48}$$

The result has a fascinating geometrical interpretation. From (6.40)

$$f_B(\vec{0}) = |\det \mathbf{Q}| \int_B d\vec{u}$$

But, with an illustration in Figure 6.16,

$$B = \int_B d\vec{u} = \text{area of integration, } \mathcal{B}.$$

and

$$C = \int_C d\vec{u} = \text{area of a cell, } C = |\det \mathbf{P}|.$$

Thus (6.48) can be written as

$$\frac{\overline{\eta^2(\vec{0})}}{\overline{\xi^2}} = \left(\frac{C}{B} - 1\right)^{-1} \tag{6.49}$$

The restoration noise level is thus directly determined by the area of the integration region for $f_B(\vec{t})$ and the area of a cell. Equation (6.49) is a strictly increasing function of B. Thus, for minimum restoration noise level, we choose $B = A =$ the region of support of the signal $x(\vec{t})$.

For Nyquist density sampling in one dimension, $A = B = C$. In this case oversampling is required to restore lost samples. For higher dimensions, the restoration capability is dependent on the region of support of the signal's spectrum. If the support is in the shape of a cell (e.g., rectangular, hexagonal), then restoration is not possible at the Nyquist density.

6.3.2.1   Filtering

Discrete white noise has a uniform spectral density and thus significant high-frequency energy. Once lost data have been restored, the data noise level can be reduced by filtering the result through B assuming that $B < C$. The noise level at the lost sample location remains the same. The noise level at locations far removed from the lost-sample locations will asymptotically be the same as that for the filtered noisy samples if no data were lost. If $\xi(\mathbf{Q}\vec{n})$ is zero mean and stationary, then after filtering, the process $\Psi(\mathbf{Q}\vec{n})$ is also stationary. If the data noise is white as in (6.46), its spectral density is uniform in C. Thus if we filter the noise through $\mathcal{B}$, the resulting normalized noise level is

$$\frac{\overline{\Psi^2}}{\overline{\xi^2}} = \frac{B}{C} \tag{6.50}$$

(A more rigorous derivation is left as an exercise.) To minimize, we clearly would choose $B = A$.

For a single lost sample in discrete white noise, the ratio of the restoration noise level to that of data far removed is, after filtering through $\mathcal{B}$,

$$\frac{\overline{\eta^2(\vec{0})}}{\overline{\Psi^2}} = \left(1 - \frac{B}{C}\right)^{-1} \tag{6.51}$$

where we have used (6.49) and (6.50). To minimize, we again would choose $B = A$. Note that (6.51) exceeds both unity and (6.49).

6.3.2.2   Deleting Samples from Optical Images

The Nyquist sampling density for images whose spectra have circular support is achieved when the circles in the frequency domain are densely packed as is shown at the top of Figure 6.13.

Note, as is shown at the bottom of Figure 6.13, that the area of $\mathcal{A}$ is less than that of C. Thus, in the absence of noise, an arbitrary number of lost image samples can be restored from those (infinite number) remaining. For $B = A$, the interpolation function here is

$$f_A(t_1, t_2) = 2W^2 |\det \mathbf{Q}| \text{jinc}(Wr) = \frac{\text{jinc}(Wr)}{\sqrt{3}}$$

We can numerically illustrate the effects of discrete white noise on restoring a lost sample from an image that has a spectrum with circular support. Suboptimal rectangular sampling is considered first, followed by the optimal hexagonal case. Both cases are extended to higher dimensions.

Rectangular Sampling

If limited to rectangular sampling, (see Figure 6.13), the restoration noise level from (6.49) follows as

$$\frac{\overline{\eta^2(\vec{0})}}{\overline{\xi^2}} = \left(\frac{4}{\pi} - 1\right)^{-1} \approx 3.66 \qquad (6.52)$$

After filtering through the $\mathcal{A}$ circle, the ratio of the restoration noise level to the data noise level at points far removed from the origin is

$$\frac{\overline{\eta^2(\vec{0})}}{\overline{\Psi^2}} = \left(1 - \frac{\pi}{4}\right)^{-1} \approx 4.66 \qquad (6.53)$$

where we have used (6.51) with $B = A = \pi W^2$. The lost-sample noise is thus 6.7 dB above the filtered data noise at infinity.

The results can easily be extended to higher dimensions. Assume that the spectrum has support within an N-dimensional hypersphere of radius W [Wozencraft and Jacobs]:

$$A = \begin{cases} \dfrac{2^N \pi^{\frac{N-1}{2}} \left(\frac{N-1}{2}\right)! W^N}{N!} & ; \quad \text{odd } N \\ \dfrac{\pi^{\frac{N}{2}} W^N}{\left(\frac{N}{2}\right)!} & ; \quad \text{even } N \end{cases} \qquad (6.54)$$

For rectangular sampling, $C = (2W)^N$. The corresponding plots of $\overline{\eta^2(\vec{0})}/\overline{\xi^2}$ and $\overline{\eta^2(\vec{0})}/\overline{\Psi^2}$ are shown as solid lines in Figure 6.18.

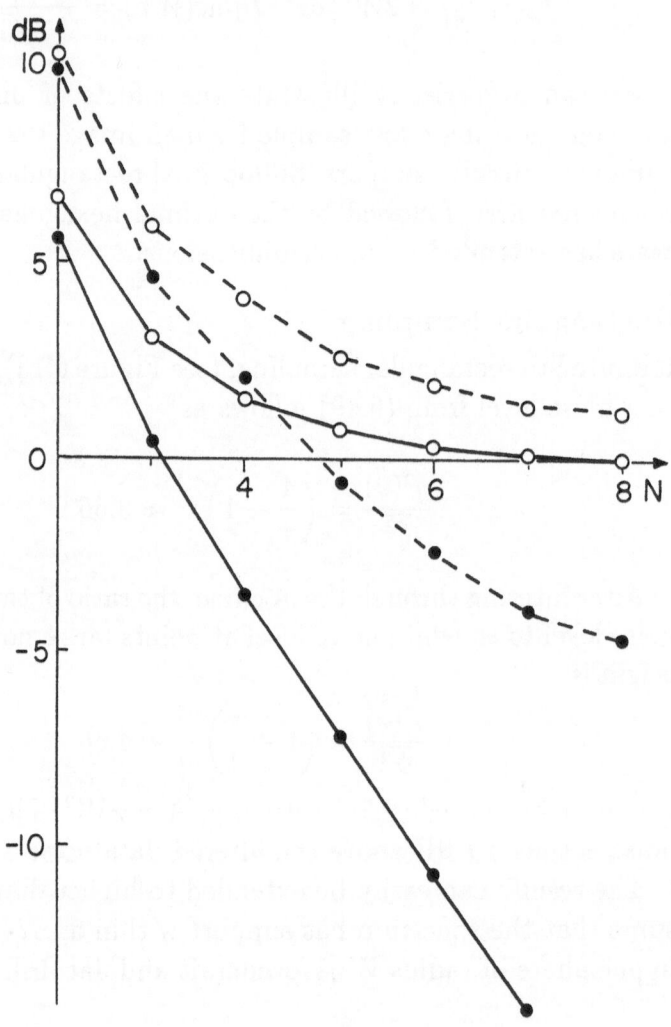

Figure 6.18: Plots of $\overline{\eta^2}/\overline{\xi^2}$ (filled circles) and $\overline{\eta^2}/\overline{\Psi^2}$ (open circles) in dB $[10\log_{10}(\cdot)]$. The solid lines are for minimum density rectangular sampling and the dashed for Nyquist (hexagonal) sampling.

Hexagonal Sampling

A single hexagonal cell is shown at the bottom of Figure 6.14 for minimum density sampling. The area of the hexagon is

$$C = 2\sqrt{3}W^2$$

Thus, from (6.49) for $B = A = \pi W^2$

$$\frac{\overline{\eta^2(\vec{0})}}{\overline{\xi^2}} = \left[\frac{2\sqrt{3}}{\pi} - 1\right]^{-1} \approx 9.74$$

and, similarly, from (6.51)

$$\frac{\overline{\eta^2(\vec{0})}}{\overline{\Psi^2}} = \left[1 - \frac{\pi}{2\sqrt{3}}\right]^{-1} \approx 10.74$$

As one would expect, these values ($\approx$ 10dB) are greater than those of the corresponding rectangular sampling cases in (6.52) and (6.53).

In higher dimensions, Nyquist sampling corresponds to densely packing hyperspheres in the frequency domain. We use Table 6.2 in conjunction with (6.54) to generate the restoration noise level plots in Figure 6.18 for Nyquist density sampling when the signal's spectrum support is a hypersphere. The plots are shown with broken lines and, as we would expect, exceed the corresponding rectangular sampling results.

6.4   Periodic Sample Decimation and Restoration

In the previous section, we showed that if gaps exist in the replication of spectra, then an arbitrarily large but finite number of lost samples can be regained from those remaining. In this section, we show that under the same circumstances, certain periodic sample decimation can be reversed from knowledge of the remaining samples. Thus, an infinite number of lost samples can be restored in certain scenarios. This procedure is applicable in certain cases even at the Nyquist density. The overall sampling density can thus be reduced to below the Nyquist density.

Fundamentally, we will distinguish between the Nyquist and *minimum sampling* densities which, in one dimension, are the same. The Nyquist density is the area of a periodicity cell when the replicated spectra are most densely packed. The minimum sampling density, we will show, is equal to the area of the support of a signal's spectrum. With reference to Figure 6.14, we showed, for example, that the Nyquist density for a signal with a circular spectrum is, from (6.38), $2\sqrt{3}W^2$. We will show that the minimum sampling density is the area of the circle, πW^2, a sampling density reduction of over 9%. In practice, this reduction can achieved by the decimation procedure described in this section.

6.4.1   Preliminaries

Before a discussion of the procedure to restore decimated samples, we need to establish a formality for decimation notation. Consider sampling geometry at the top of Figure 6.19 with sampling matrix

$$\mathbf{Q} = \begin{bmatrix} T & -T \\ T & T \end{bmatrix}$$

As shown, the samples are divided into four groups labeled one through four. The sampling matrix for the group of solid dots is clearly

$$\mathbf{D} = \begin{bmatrix} 2T & -2T \\ 2T & 2T \end{bmatrix}$$

Each of the other subgroups has the same sampling matrix, but with a different offset vector. Using the four samples in the bold square diamond, these offset vectors are

$$\mathbf{Q}\vec{e}_1 = \begin{bmatrix} T \\ -T \end{bmatrix}$$

$$\mathbf{Q}\vec{e}_2 = \begin{bmatrix} 0 \\ 0 \end{bmatrix}$$

$$\mathbf{Q}\vec{e}_3 = \begin{bmatrix} -T \\ -T \end{bmatrix}$$

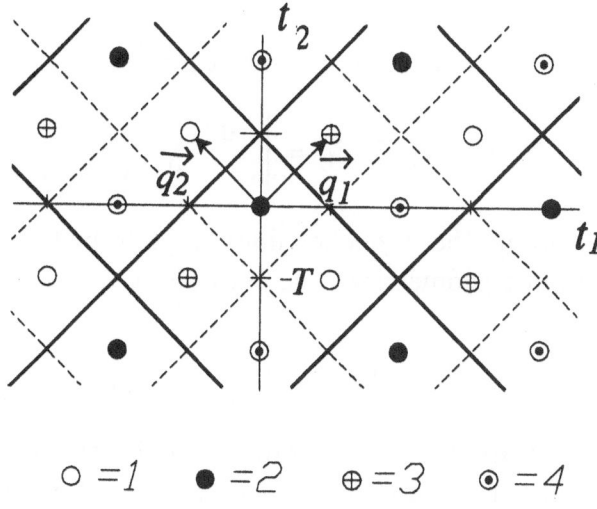

$$\circ = 1 \quad \bullet = 2 \quad \oplus = 3 \quad \odot = 4$$

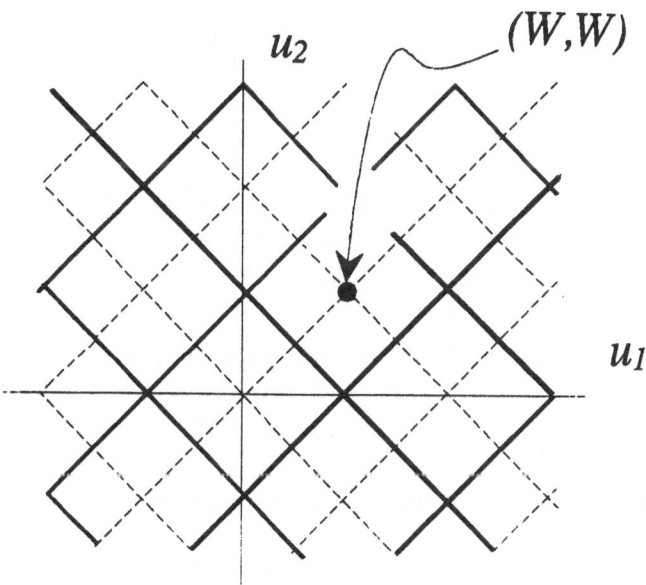

Figure 6.19: (Top) Samples are divided into four subgroups labeled 1 through 4. (Bottom) A periodicity cell for all the samples is the large square diamond. Each diamond contains four subcells. A subcell is a periodicity cell for any one sample subgroup.

$$Q\vec{e}_4 = \begin{bmatrix} 0 \\ -2T \end{bmatrix}$$

Correspondingly, we define the offset matrix

$$E = [\vec{e}_1, \vec{e}_2, \vec{e}_3, \vec{e}_4] = \begin{bmatrix} 0 & 0 & -1 & -1 \\ -1 & 0 & 0 & -1 \end{bmatrix}$$

Let's examine the frequency domain periodicity generated by these sampling geometries. The periodicity matrix for all of the samples is

$$P = Q^{-T} = \begin{bmatrix} W & -W \\ W & W \end{bmatrix}$$

The large square diamonds shown at the bottom of Figure 6.19 are corresponding periodicity cells. (Each contains four smaller square diamonds.) For any one of the sample subgroups, the periodicity matrix is

$$P_D = D^{-T} = \begin{bmatrix} \frac{W}{2} & -\frac{W}{2} \\ \frac{W}{2} & \frac{W}{2} \end{bmatrix}$$

The smaller square diamonds at the bottom of Figure 6.19 outlined by broken lines can serve as periodicity *subcells* for a sample subgroup, A cell, C, and subcell, C_D, for this example are shown in Figure 6.20.

We can generalize our observations. For a given Q, we have

$$D = QM$$

where M is a nonsingular matrix of integers. For our previous example

$$M = \begin{bmatrix} 2 & 0 \\ 0 & 2 \end{bmatrix}$$

The E matrix of offset vectors is obtained through examination of any period of sample subgroups. There are

$$L = |\det M|$$

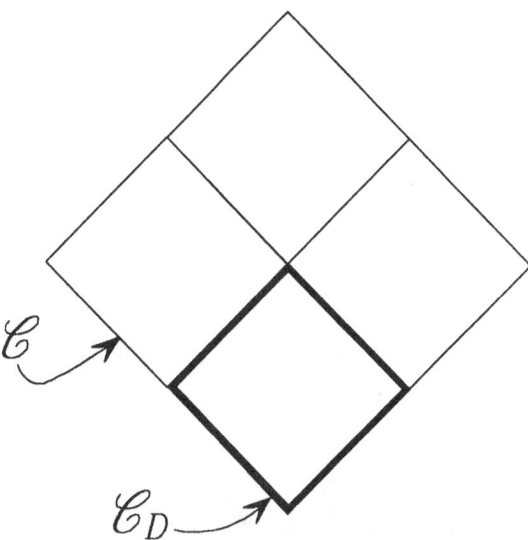

Figure 6.20: A periodicity cell divided into four subcells.

subgroups and L subcells per periodicity cell.

The spectrum of all of the samples is

$$s(\vec{t}) = \sum_{\vec{n}} x(\mathbf{Q}\vec{n})\delta(\vec{t} - \mathbf{Q}\vec{n})$$

$$\leftrightarrow S(\vec{u}) = \sum_{\vec{n}} x(\mathbf{Q}\vec{n})\exp(-j2\pi\vec{u}^T\mathbf{Q}\vec{n})$$

The samples can be recovered via

$$x(\mathbf{Q}\vec{n}) = |\det \mathbf{M}| \int_{\mathcal{C}} S(\vec{u})\exp(j2\pi\vec{u}^T\mathbf{Q}\vec{n})d\vec{u}$$

where $\mathcal{C}$ is any periodicity cell.

The ith sample subgroup has a spectrum of

$$s_i(\vec{t}) = \sum_{\vec{n}} x(\mathbf{D}\vec{n} - \mathbf{Q}\vec{e}_i)\delta(\vec{t} - \mathbf{D}\vec{n} + \mathbf{Q}\vec{e}_i) \leftrightarrow$$

$$S_i(\vec{u}) = \exp(-j2\pi\vec{u}^T\mathbf{Q}\vec{e}_i) \ \sum_{\vec{n}} x(\mathbf{D}\vec{n} - \mathbf{Q}\vec{e}_i)\exp(-j2\pi\vec{u}^T\mathbf{D}\vec{n});$$

$$1 \le i \le L \qquad (6.55)$$

A sample subgroup can be obtained from

$$x(\mathbf{D}\vec{n} - \mathbf{Q}\vec{e}_i) = |\det \mathbf{D}| \int_{\mathcal{C}_D} S_i(\vec{u}) \exp[j2\pi\vec{u}^T(\mathbf{D}\vec{n} - \mathbf{Q}\vec{e}_i)]d\vec{u}$$

(6.56)

where $\mathcal{C}_D$ is any subcell. Clearly

$$S(\vec{u}) = \sum_{i=1}^{L} S_i(\vec{u})$$

(6.57)

6.4.2   First Order Decimated Sample Restoration

Consider unaliased replication of a spectrum in such a manner that gaps occur. As we have seen, this can even occur at Nyquist densities. We will show that if a subcell is totally subsumed in a gap contained within a cell, then any sample subgroup can be expressed as a linear combination of those sample subgroups remaining. Thus, a sample subgroup can be lost and the signal can still be interpolated from those samples remaining.

Proof :

If a subcell, $\mathcal{C}_D$, lies in a gap, then $S(\vec{u})$ is identically zero there. Thus, for $\vec{u}$ within the subcell we have from (6.57)

$$S(\vec{u}) = 0 = S_L(\vec{u}) + \sum_{i=1}^{L-1} S_i(\vec{u})$$

or

$$S_L(\vec{u}) = -\sum_{i=1}^{L-1} S_i(\vec{u})$$

Substituting into (6.56) and simplifying gives

$$x(\mathbf{D}\vec{n}-\mathbf{Q}\vec{e}_L) = -|\det \mathbf{D}| \sum_{i=1}^{L-1} \int_{\mathcal{C}_D} S_i(\vec{u}) \exp[j2\pi\vec{u}^T(\mathbf{D}\vec{n}-\mathbf{Q}\vec{e}_L)]d\vec{u}$$

Substituting (6.55) and simplifying gives

$$x(\mathbf{D}\vec{n} - \mathbf{Q}\vec{e}_L) = -\sum_{i=1}^{L-1}\sum_{\vec{m}} x(\mathbf{D}\vec{n} - \mathbf{Q}\vec{e}_i)f[\mathbf{D}(\vec{n} - \vec{m}) + \mathbf{Q}(\vec{e}_i - \vec{e}_L)]$$

(6.58)

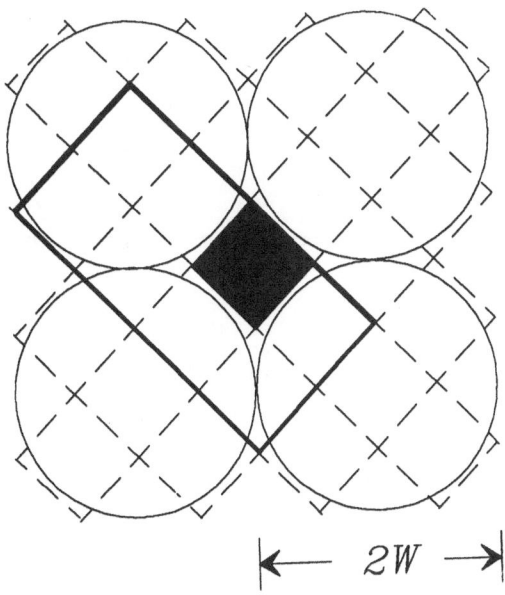

$$\longleftarrow 2W \longrightarrow$$

Figure 6.21: A periodicity cell for the replicated circles shown is the bold outlined tilted rectangle. The rectangle can be divided into eight square diamond subcells one of which, shown shaded, lies within a gap among the circles.

where

$$f(\vec{t}) = -|\det \mathbf{D}| \int_{\mathcal{C}_D} \exp(j2\pi \vec{u}^T \vec{t}) d\vec{u} \qquad (6.59)$$

is the interpolation kernel. Equation (6.58) shows the manner by which the Lth sample subgroup can be recovered from the remaining $L-1$ subgroups.

Example : We reconsider the minimum rectangular sampling density of $N = 2$ dimensional signals whose spectrum has a circular support. The spectral replication of Figure 6.12 is redrawn in Figure 6.21 with a bold outlined tilted rectangle shown as a periodicity cell. We divide the cell into eight identical square di-

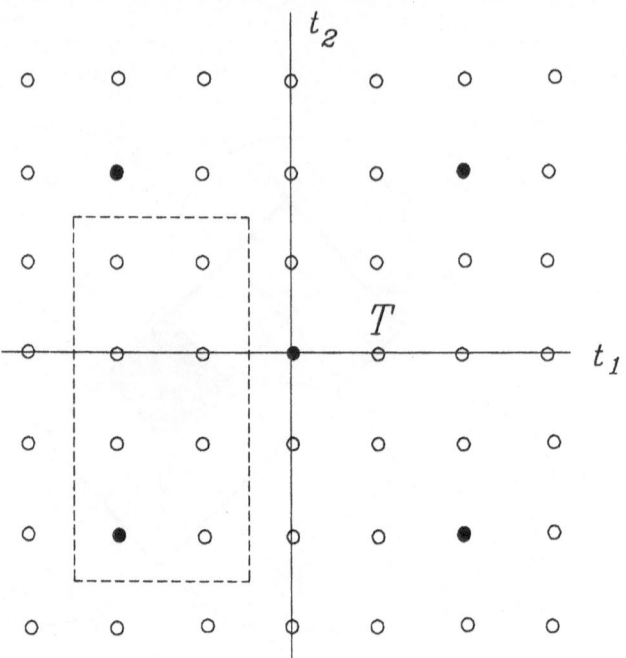

Figure 6.22: The decimation shown with solid dots achieves the periodicity subcell structure shown in Figure 6.21. Therefore, the samples at the locations of the solid dots can be deleted and restored as a linear combination of those remaining.

amond subcells. One of the cells clearly falls into the gap among replications. Using the periodicity matrix in (6.35), we find such subcells can be generated by using

$$\mathbf{M} = \begin{bmatrix} 2 & -2 \\ 2 & 2 \end{bmatrix}$$

A sample grouping that will achieve these subcells is shown in Figure 6.22. The deleted samples are shown as solid dots.

The interpolation function for this example follows from (6.59) using the square diamond in Figure 6.21 as the integration period, $\mathcal{C}_D$. The result is

$$f(t_1, t_2) = -\text{sinc}\left(\frac{t_1 + t_2}{4T}\right) \text{sinc}\left(\frac{t_1 - t_2}{4T}\right) e^{\frac{j\pi(t_1 + t_2)}{T}}$$

This decimation geometry reduces the sampling rate to $\frac{7}{8}(2W)^2$ $= 3.50W^2$ which is still higher than the Nyquist density of $2\sqrt{3}W^2 \approx 3.46W^2$.

6.4.3   Sampling Below the Nyquist Density

If there are gaps among spectral replications at the Nyquist density, then first order sample decimations can always be applied thereby reducing the overall sampling density below that of Nyquist [Cheung] [Cheung and Marks]. We offer two examples.

Example 1 : A spectral support is shown in Figure 6.23 that is zero inside the small square and zero outside the large square. The Nyquist density is clearly $(2W)^2$. The large square, however, can be divided into a three by three array of smaller squares, the center one of which falls within an identically zero region. We may therefore decimate every ninth sample as shown in Figure 6.24. The decimated samples can be recovered using (6.59) with $\mathcal{C}_D$ as the small square in Figure 6.23. The required interpolation function follows as

$$f(t_1, t_2) = -\text{sinc}\left(\frac{2Wt_1}{3}\right) \text{sinc}\left(\frac{2Wt_2}{3}\right)$$

The resulting sampling density is 8/9 that of Nyquist and is equal to the area of the spectral support.

Example 2 - Optical Images : We return to the example of maximally packed circles as illustrated in Figure 6.13. A periodicity cell for this replication follows from choosing midpoints at four gaps and forming the parallelogram shown in Figure 6.25.

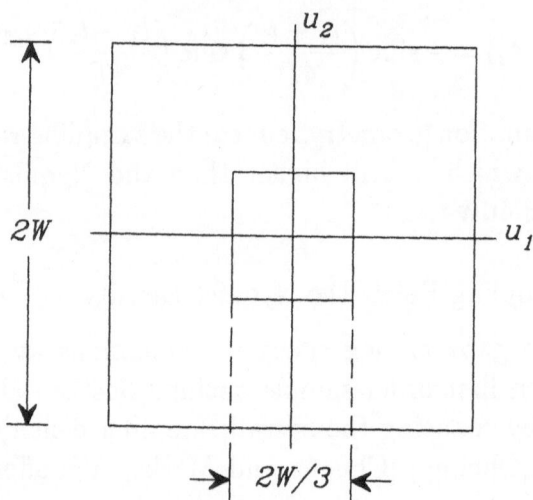

Figure 6.23: A spectral support for which a sub-Nyquist sampling density is possible.

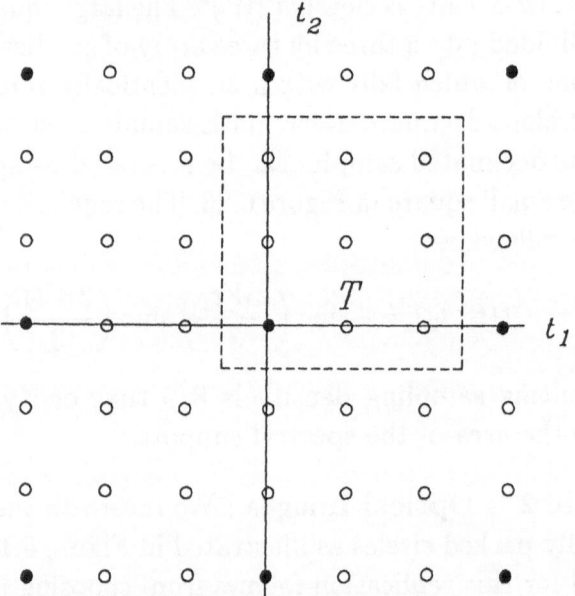

Figure 6.24: If a signal has the spectral support shown in Figure 6.23, then every ninth sample can be decimated as shown here. The composite sampling density is below that of Nyquist.

The cell is divided into four congruent parallelograms. The upper right parallelogram is divided into four smaller parallelograms. The process is repeated one final time. A small parallelogram, shown shaded, is totally contained in a gap. Using this as a subcell, we see that we can reduce the overall sampling density to $\frac{63}{64}$ that of Nyquist. A sample decimation procedure to achieve this is shown in Figure 6.26.

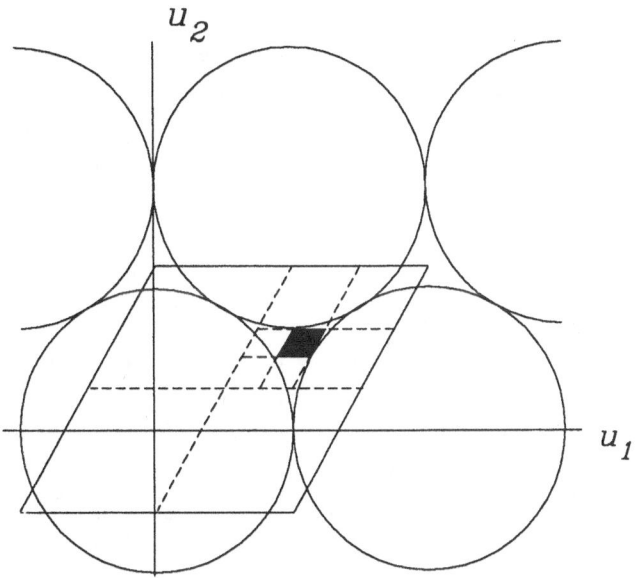

Figure 6.25: The small shaded parallelogram subcell lies totally in the gap among maximally packed circles. Since the subcell is $\frac{1}{64}^{th}$ the area of the larger parallelogram cell, the (Nyquist) sampling density can be reduced by a factor of $\frac{63}{64}$.

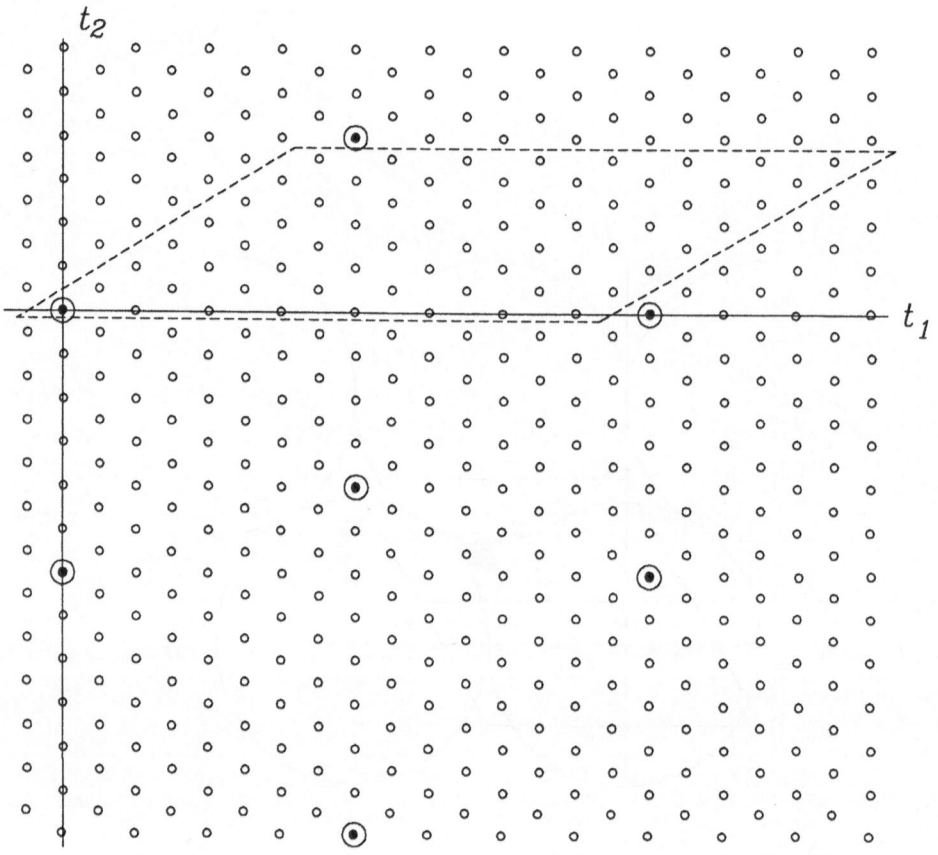

Figure 6.26: A sampling decimation procedure that will achieve the subcell structure shown in Figure 6.25.

6.4.4   Higher Order Decimation

First order decimation restoration can be straightforwardly generalized. If $M < L$ nonoverlapping subcells lie in a gap within one periodicity cell, then M sample subgroups can be represented as linear combinations of the remaining $L - M$. The analysis procedure is identical to that in Sec. 6.4.2., except that one solves for the M sample subgroups' spectra corresponding to M functional equations. The analysis is similar to a multidimensional extension of Papoulis' Generalization as presented in Chapter 4. Details are given elsewhere [Cheung], [Cheung and Marks].

Higher order decimation can be used to establish that the minimum sampling density for a bandlimited signal is equal to the area of its support. In Figure 6.27, for example, we have replication of circular support resulting from rectangular sampling.

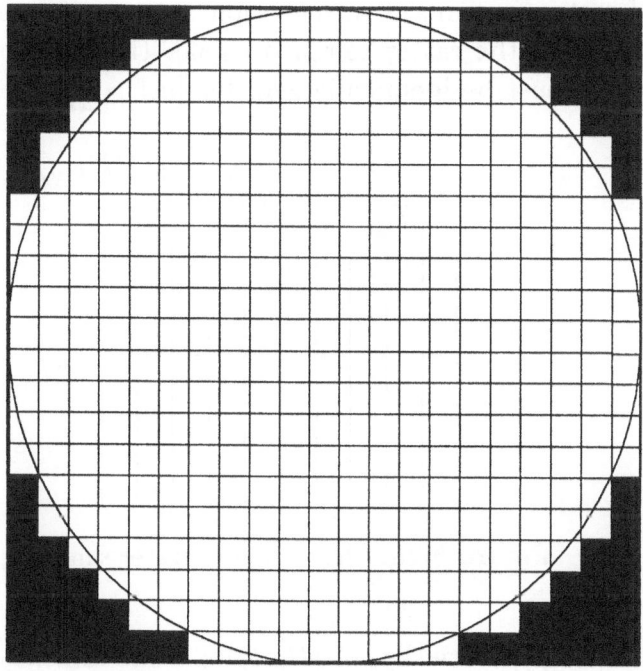

Figure 6.27: Higher order decimation can be used to illustrate that the sampling density for a bandlimited image can be reduced to the area of its support.

Each large square periodicity cell is divided into 441 square sub-cells. In each cell, 68 of these subcells (shown shaded) lie in a gap. Thus, if the circular radius is W, the sampling rate is reduced from $(2W)^2$ to $\frac{373}{441}(2W)^2 = 3.38W^2$, which is below the Nyquist density of $3.46W^2$. Ultimately, by increasing the number of subcells, the sampling density can be reduced to $3.46W^2$, the area of the circular support.

6.5   Raster Sampling

Raster Sampling of a two dimensional signal, $x(t_1, t_2)$, is illustrated in Figure 6.28. The one dimensional vertical slices can be expressed via

$$s(t_1, t_2) = \sum_{n=-\infty}^{\infty} x(nT, t_2)\delta(t_1 - nT)$$

where T is the sampling interval. The most familiar application of raster scanning is in television. In order for there to be no aliasing in the raster sampled signal, the signal spectrum, $X(u_1, u_2)$, must be identically zero for $|u_2| > \frac{1}{2T}$ for all values of u_1.

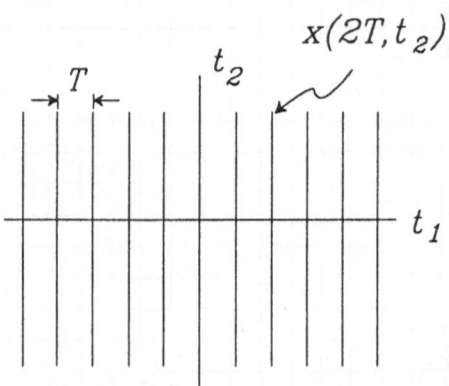

Figure 6.28: Illustration of raster sampling.

Interestingly, every slice of $x(t_1, t_2)$ is a one-dimensional function with the same spectral support (*e.g.* bandwidth). We can establish this by inspection of Figure 6.29 where the support of a bandlimited signal, $X(u_1, u_2)$, is shown in the bottom right corner. We can obtain $x(t_1, t_2)$ by first inverse transforming $X(u_1, u_2)$ in the u_2 direction and then the u_1 direction. The result of the first step is pictured in the upper right corner of Figure 6.29. The vertical slice of this function at $u_1 = v$ is simply the inverse transform of the slice $X(v, u_2)$. Since, for a fixed v, this is a bandlimited function, it cannot be zero on any finite interval. Extending this to all of the vertical slices of $X(u_1, u_2)$, we conclude that the one dimensional inverse transform of $X(u_1, u_2)$ pictured in the upper right hand corner is identically equal to zero outside of the strip $|u_1| > B$ and is almost everywhere nonzero inside the strip.

Consider, then, the one dimensional slice, $x(t_1, \tau)$, shown at the upper left of Figure 6.29. For a given τ, what is the bandwidth of this signal? Taking the one dimensional Fourier transform results in the $t_2 = \tau$ slice shown at the upper right of Figure 6.29. From our previous arguments, the bandwidth of $x(\tau, t_1)$ is B. Indeed, the bandwidth of all the horizontal slices of $x(t_1, t_2)$ is B. We therefore conclude that *any two parallel slices of a two dimensional bandlimited function have the same spectral support (e.g. bandwidth).*

The required raster sampling rate, $2W$, is clearly deduced by inspection of the lower left corner of Figure 6.29. This rate can be finite even when B is not. The $2W$ support interval is the geometric shadow of the support of $X(u_1, u_2)$. This rate will generally vary as we rotate the signal and thereby rotate its spectrum.

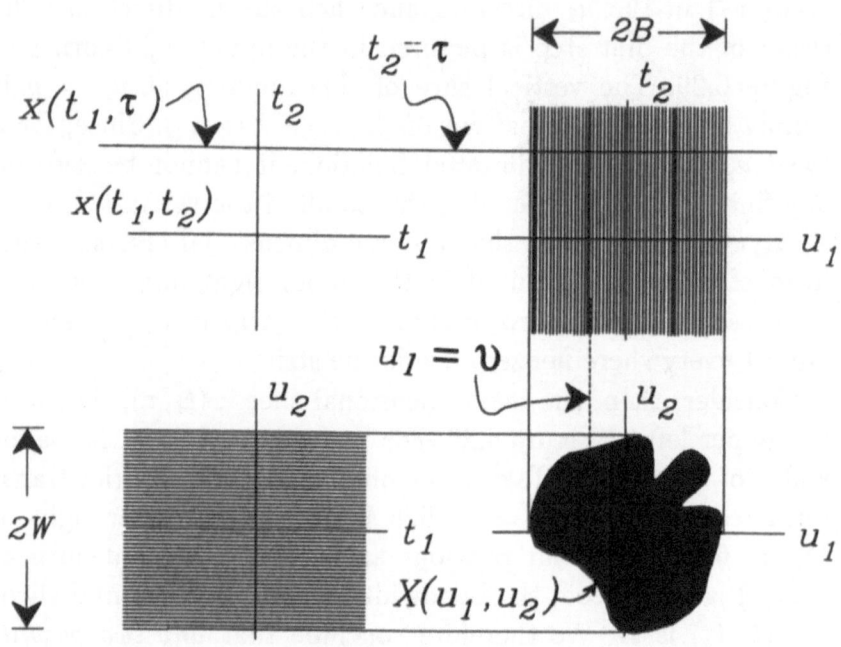

Figure 6.29: A two dimensional signal can be Fourier transformed by first transforming each horizontal slice (upper right hand corner) and then Fourier transforming each vertical slice of the resulting function. Alternately, we can first Fourier transform each vertical slice (lower left hand corner)

6.6 Exercises

6.1 Compute the Fourier transform of the two dimensional functions shown in Figure 6.30. All are one inside and zero outside the curves shown. The ellipses in (c) and (d) are identical.

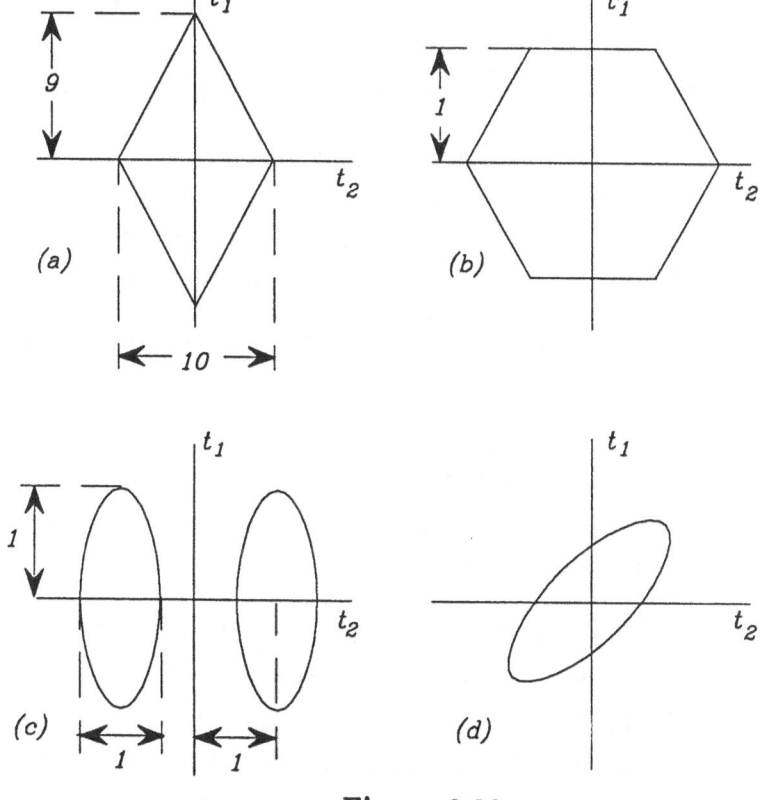

Figure 6.30

6.2 Compute the Hankel transform of $\exp(-\pi r^2)$.

6.3 Let $f(r) \leftrightarrow F(\rho)$ denote Hankel transform pairs. Complete the following theorems:

(a) Scaling

$$f(ar) \leftrightarrow?$$

(b) Inversion

$$f(r) = 2\pi \int_0^\infty \rho F(\rho) J_0(2\pi r \rho) d\rho$$

6.4 Show that the magnitudes of the circular harmonics of $x(r, \theta)$ are the same as those for the rotated image $x(r, \theta - \phi)$.

6.5 Compute the Fourier series coefficients for the following periodic functions

(a) $S(\vec{u}) = \sum_{\vec{n}} G(\vec{u} - \mathbf{P}\vec{n})$ where $G(\vec{u})$ is a given function that is <u>not</u> a period of $S(\vec{u})$

(b) $S(u_1, u_2) = \cos[2\pi(\sin u_1 + \sin u_2)]$

6.6 A function, $x(t_1, t_2)$, is sampled using the sampling matrix in (6.36). All samples are zero except for $x(0,0) = 1$. Assuming no aliasing and a cell equal to the tilted rectangle in Figure 6.10, give a numerical value for $x(3,0)$.

6.7 A sampling geometry is denoted by $\mathbf{Q}$. Show that $\mathbf{QM}$ produces the same geometry when $\mathbf{M}$ is a matrix of integers and $|\det \mathbf{M}| = 1$.

6.8 Replace the sample data in (6.39) by the noisy data $x(\mathbf{Q}\vec{m})$ $+\xi(\mathbf{Q}\vec{m})$. The result is $x(\vec{t}) + \Psi(\vec{t})$. Assume that $\xi(\mathbf{Q}\vec{m})$ is zero mean and wide sense stationary with variance $\overline{\xi^2}$.

(a) Show for $B = C$, that $\overline{\Psi^2(\vec{t})} = \overline{\xi^2}$. Thus the interpolation noise level is the same as the data noise level.

(b) Show that when the data noise is discrete white noise, that the interpolation noise level is given by (6.50).

6.9 A signal $x(t_1, t_2)$ is known to have a spectrum that is identically equal to zero outside of an equilateral triangle. If a side of the triangle is B, what is a sampling matrix that we can use to minimize the sampling density? Draw the replicated spectra resulting from this sampling geometry.

6.10 A signal $x(t_1, t_2)$ is known to have a spectrum that is identically zero outside of a circle of finite radius in the (u_1, u_2) plane. The signal is sampled at its Nyquist density corresponding to this circle and the sample at the origin is lost.

If all of the known samples are zero, what is the value of the lost sample?

6.11 For a fixed $x(\vec{t})$ and $\mathbf{Q}$, numerically investigate the convergence rate of (6.44) for various B's.

6.12 Derive the interpolation function for the second example in section 6.4.3 as illustrated in Figure 6.25 and Figure 6.26.

6.13 Instead of $x(\mathbf{D}\vec{m} - \mathbf{Q}\vec{e}_i)$ on the right side of (6.58), suppose we had $x(\mathbf{D}\vec{m} - \mathbf{Q}\vec{e}_i) + \xi(\mathbf{D}\vec{m} - \mathbf{Q}\vec{e}_i)$ where ξ is zero mean discrete white noise with variance $\overline{\xi^2}$. Thus

$$E[\xi(\mathbf{D}\vec{m} - \mathbf{Q}\vec{e}_i)\xi(\mathbf{D}\vec{n} - \mathbf{Q}\vec{e}_i)] = \overline{\xi^2}\delta[\vec{n} - \vec{m}]\delta[i - j]$$

Let the response to this noisy data be

$$x(\mathbf{D}\vec{m} - \mathbf{Q}\vec{e}_L) + \eta(\vec{n})$$

Clearly, $\eta(\vec{n})$ is zero mean and has the same variance,

$$E[|\eta(\vec{n})|^2] = \overline{\eta^2}$$

for all $\vec{n}$. Find a closed form expression for the NINV, $\overline{\eta^2}/\overline{\xi^2}$.

6.14 Consider the parallelogram periodicity cell for maximally packed circles pictured in Figure 6.31. The parallelogram can be divided into two equilateral triangles. One triangle is positioned, as shown, with its three vertices centered in gaps. The triangle is divided into four smaller equilateral triangles. The center triangle is again divided. Note that the small equilateral triangle falls completely in a gap.

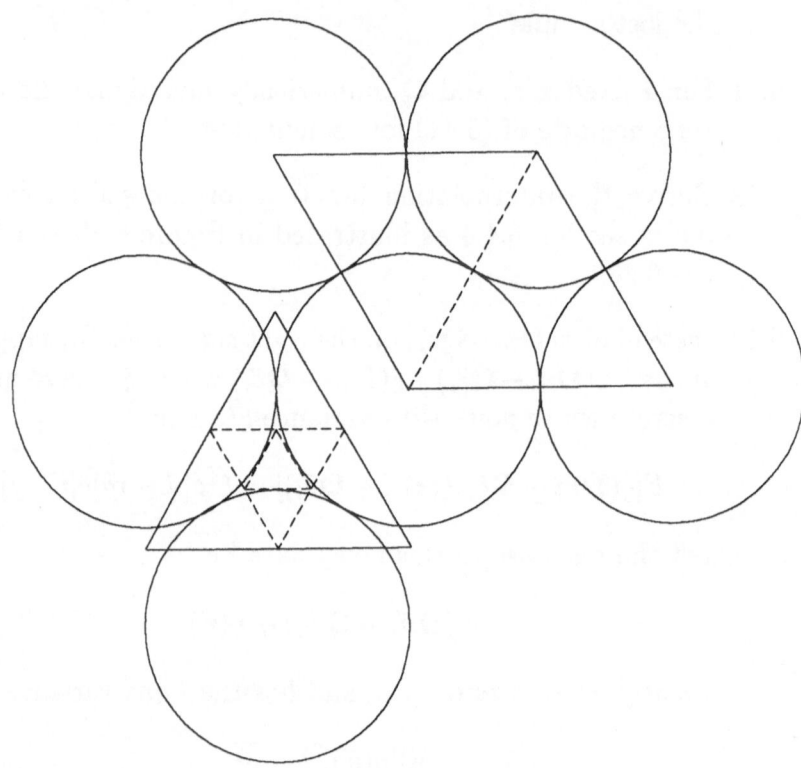

Figure 6.31

(a) Using this observation, describe a second order decimation scheme to reduce sampling below the Nyquist
rate. (Note: as shown, the periodicity cell can be ori-
ented as to contain these two gaps.)

(b) What is the overall density of this decimation scheme?

(c) For your choice of subgroup decimation, derive the
required interpolation functions.

6.15 Assume that functions in Figure 6.30 are functions of
(u_1, u_2) rather than (t_1, t_2). We wish to raster sample $N =$
2 dimensional signals that have spectra with supports illustrated in Figure 6.30(a), (b) and (d). We have the freedom to choose the sense of sampling. All sample slices, for
example, can be horizontal, all can be at $45°$ or $30°$, etc.

(a) In each case, specify a sampling sense that produces the minimum number of line samples per unit interval. Also, give this minimum rate.

(b) Sketch the resulting spectral replication.

REFERENCES

R.N. Bracewell, **The Fourier Transform and Its Applications, 2nd edition**, Revised, McGraw-Hill, NY, 1986.

K.F. Cheung, **Image Sampling Density Reduction Below that of Nyquist**, PhD Dissertation, University of Washington, Seattle (1988).

K.F. Cheung and R.J. Marks II "Image sampling below the Nyquist density without aliasing", *Journal of the Optical Society of America A*, vol.7, pp.92-105 (1990).

D.E. Dudgeon and R.M. Mersereau, **Multidimensional Digital Signal Processing**, Prentice-Hall, New Jersey, 1984.

J.D. Gaskill, **Linear Systems Fourier Transforms and Optics**, John Wiley & Sons, Inc., New York, 1978.

J.W. Goodman, **Introduction to Fourier Optics**, McGraw-Hill, New York, 1968.

R.J. Marks II "Multidimensional signal sample dependency at Nyquist densities", *Journal of the Optical Society of America A*, vol. 3, pp.268-273 (1986).

D.P. Petersen and D. Middleton "Sampling and reconstruction of wave number-limited functions in N-dimensional Euclidean spaces", *Informat. Contr*, vol. 5, pp.279-323 (1962).

J.M. Wozencraft and I.M. Jacobs, **Principles of Communication Engineering**, John Wiley & Sons,Inc., New York, 1965.

7

Continuous Sampling

To this point, sampling has been discrete. In this chapter we consider reconstruction of bandlimited signals from continuous sampling. Our definition of continuous sampling is best presented by illustration. A signal, $f(t)$, is shown in Fig. 7.1a, along with some possible continuous samples. Regaining $f(t)$ from knowledge of $g_e(t) = f(t)\Pi(t/T)$ in Fig. 7.1b is the extrapolation problem which has applications in a number of fields. In optics, for example, extrapolation in the frequency domain is termed *super resolution*.

Reconstructing $f(t)$ from its tails [*i.e.*, $g_i(t) = f(t)\{1-\Pi(t/T)\}$] is the *interval interpolation problem. Prediction*, shown in Fig. 7.1d, is the problem of recovering a signal with knowledge of that signal only for negative time.

Lastly, as illustrated in Fig. 7.1e, is *periodic continuous sampling*. Here, the signal is known in sections periodically spaced at intervals of T. The duty cycle is α. Reconstruction of $f(t)$ from this data includes a number of important reconstruction problems as special cases:

1. Assume $f(t)$ goes to zero as $t \to \pm\infty$. Then, by keeping αT constant, we can approach the extrapolation problem by letting T go to ∞.

2. Redefine the origin to be centered in a zero interval. Under the same assumption as (1), we can similarly approach the interpolation problem.

3. Redefine the origin as in Exercise 7.1. Then the interpolation problem can be solved by discarding data in Fig. 7.1c to make it periodically sampled.

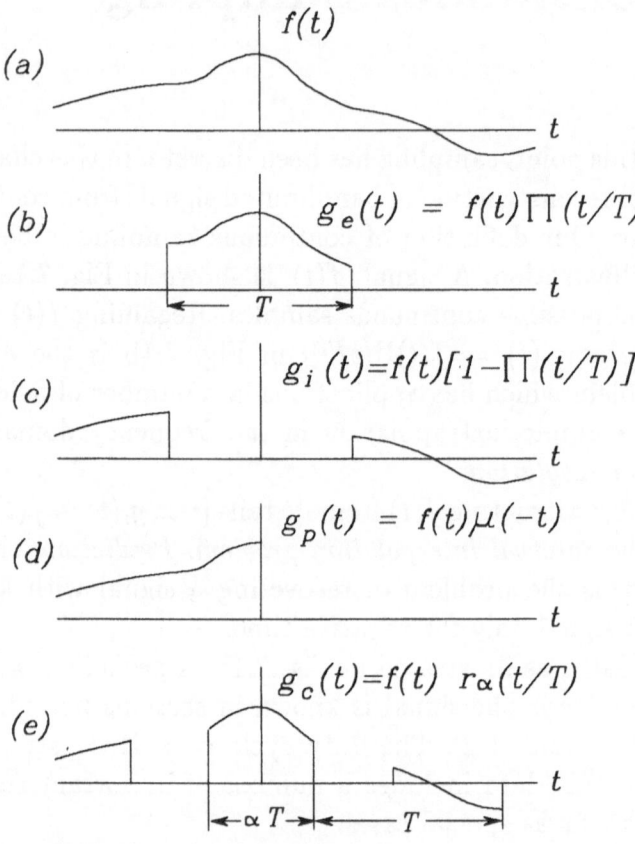

Figure 7.1: Illustration of continuous sampling (a) the original
signal; (b) extrapolation; (c) interpolation; (d) prediction; (e) periodic continuous sampling. The value
and meaning of T varies in each case.

4. Keep T constant and let $\alpha \to 0$. The result is reconstructing $f(t)$ from discrete samples as discussed in Chapter 3. Indeed, this model has been used to derive the sampling theorem [Carlson].

Figures 7.1b-e all illustrate continuously sampled versions of $f(t)$. In this chapter, we will present techniques by which the signal can be reconstructed assuming that $f(t)$ is bandlimited. In the absence of noise, any finite energy bandlimited signal can be reconstructed from any continuous sample. Such signals are *analytic* (or *entire*) everywhere [Boas, Papoulis (1977)]. Thus, if we know the signal within an arbitrarily small neighborhood centered at $t = \tau$, we can compute the value of the function and all its derivatives at τ and generate a Taylor series about $t = \tau$ that converges everywhere:

$$f(t) = \sum_{n=-\infty}^{\infty} (t - \tau)^n f^{(n)}(\tau)/n!.$$

In practice, of course, this can only be done to an approximation. We could, for example, empirically determine $f(\tau)$ and $f^{(1)}(\tau)$ and maybe even $f^{(2)}(\tau)$. But, as we have seen in Section 5.1.4., higher order derivative determination will become critically muddled by measurement inexactitude. Thus, in our example, we could at best fit a quadratic to the signal at $t = \tau$. In the absence of uncertainty, however, restoration can be performed from any continuously sampled bandlimited signal.

Intuitively, one should not be surprised that a known portion of a bandlimited signal can be extended at least near to where the signal is known. Bandlimited signals are smooth. The extension of such a signal must be similarly smooth. Simply continuing the curve will in general yield a good estimate "near" to where the signal is known.

For the extrapolation problem, Pask explained the relationship between the known interval and those samples far removed from the interval using the cardinal series interpolation. In Fig. 7.2, the sample $f(n/2W)$ is assumed to be outside of the known interval. From the sampling theorem, the interpolation contribution of this sample is a sinc function whose tails will intersect

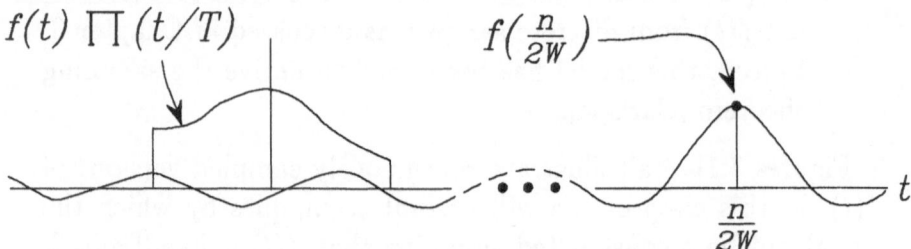

Figure 7.2: Illustration that a sample value far removed from an
interval has an effect on the signal in that interval.

and thus make a contribution to the known portion of the signal. Thus, the known portion of the signal contains information about the unknown part of the signal. Note, however, as we go farther and farther away from the known portion of the signal, the contribution becomes less and less.

Unlike the Taylor series treatment, most of the restoration algorithms in this chapter make use of the entire known portion of the signal. Like the Taylor series treatment, the sensitivity of these algorithms to inexactitude must be considered. Indeed, a number of the algorithms are *ill-posed*. This means that a small amount of noise on the known data can render the reconstruction unstable. In such cases, further *a priori* information about $f(t)$ must be included in the algorithm *i.e.*, the original problem statement is too vague. If we assume only that $f(t)$ is bandlimited with finite energy, the extrapolation and prediction problems are ill-posed whereas restoration from periodic continuous samples and interpolation problems are not.

7.1 Interpolation From Periodic Continuous Samples

The given data for the periodic continuously sampled signal in Fig. 7.1e is

$$g_c(t) = f(t) r_\alpha(\frac{t}{T}) \qquad (7.1)$$

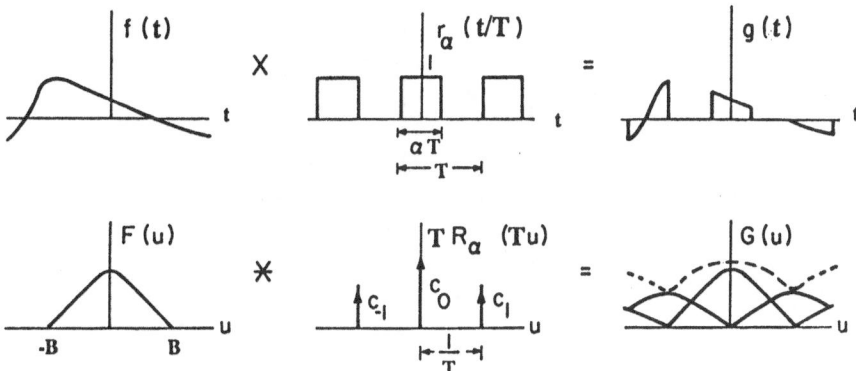

Figure 7.3: Illustration of the degradation of $f(t)$ to $g(t)$ (a) in t. (b) in the frequency domain.

where the duty cycle, α, lies between zero and one and the pulse train is

$$r_\alpha(t) = \sum_{n=-\infty}^{\infty} \Pi(\frac{t-n}{\alpha}). \tag{7.2}$$

Given that $f(t)$ is bandlimited, the problem is to find $f(t)$ given $g_c(t)$ and the signal's bandwidth, B.

The degradation process described by (7.1) is illustrated by the top three functions in Fig. 7.3. The corresponding operation in the frequency domain, shown in the bottom three functions in Fig. 7.3 is

$$G_c(u) = F(u) * T R_\alpha(Tu) \tag{7.3}$$

where the upper case letters denote the Fourier transforms of the corresponding functions in (7.1) and the asterisk denotes convolution. Expanding (7.2) in a Fourier series followed by transformation gives

$$T R_\alpha(Tu) = \sum_{n=-\infty}^{\infty} C_n \delta(u - \tfrac{n}{T}) \tag{7.4}$$

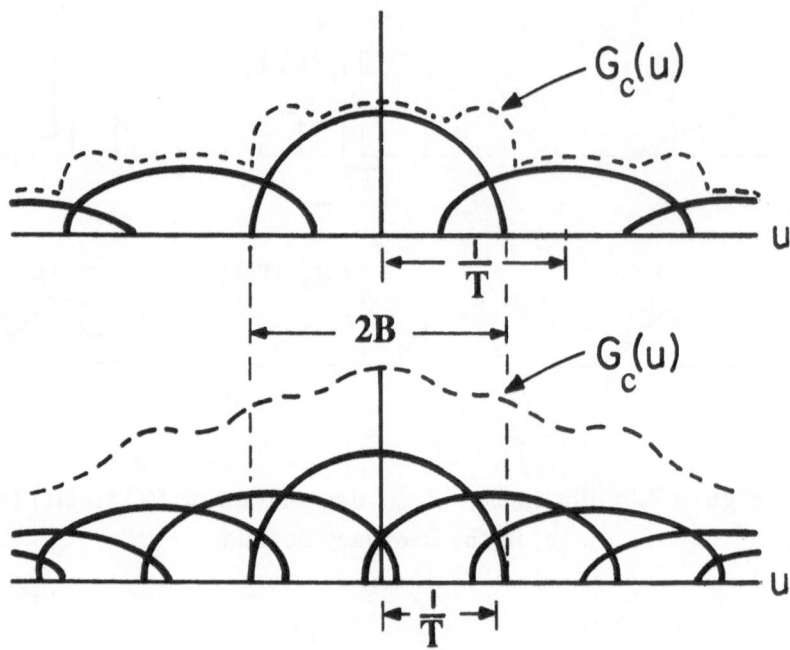

Figure 7.4: Illustration of (a) first order aliasing. (b) second order aliasing.

where
$$C_n = \alpha\mathrm{sinc}(\alpha n).$$

Then the spectrum of the degradation in (7.3) can be written:

$$G_c(u) = \sum_{n=-\infty}^{\infty} C_n F(u - \tfrac{n}{T}).$$

7.1.1 The Restoration Algorithm

Clearly, if the sampling rate $1/T$ exceeds $2B$, the replicated spectra do not overlap and $F(u)$ can be regained from $G(u)$ by a simple low pass filter. We are interested in restoration when the data is aliased. If one of the spectra overlaps the right half

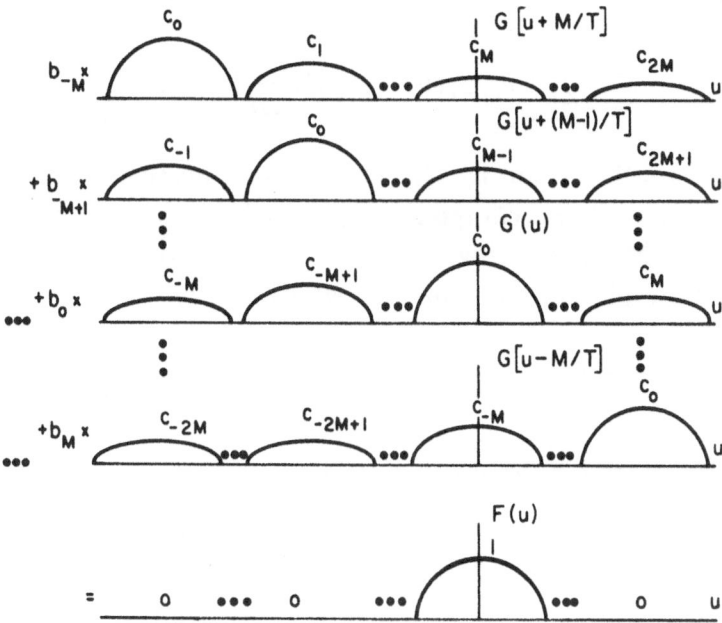

Figure 7.5: Illustration of the methodology of restoring M-th order aliased data by summing $2M + 1$ shifted and weighted versions for the degraded spectrum.

zero order spectrum as in Fig. 7.4a, we have first order aliasing. If two overlap, as in Fig. 7.4b, we have second order aliasing, etc. In general, the order of aliasing is

$$M =< 2BT >$$

where $< \zeta >$ denotes "the greatest integer not exceeding ζ."

Consider Fig. 7.5 in which $2M+1$ shifted versions of $G_c(u)$ are shown, *i.e.*, the set

$$\{G_c(u - \frac{m}{T}) \mid -M \leq m \leq M\}.$$

The interfering component spectra in each shifted G_c are shown not overlapping for presentation clarity. We now simply need to weight the mth shifted G_c by a coefficient b_m so that

$$\sum_{m=-M}^{M} b_m G_c(u - \frac{m}{T}) \, \Pi(\frac{u}{2B}) = F(u). \tag{7.5}$$

With attention again to Fig. 7.5, this is equivalent to summing the weights of the component spectra in each column to give zero for the interferring spectra and unity for the zero order spectrum. That is, find the b_m's which satisfy

$$\sum_{m=-M}^{M} b_m C_{n-m} = \delta[n] \qquad ; -M \le n \le M. \tag{7.6}$$

Viewing this as a matrix operation:

$$\begin{bmatrix} C_0 & C_{-1} & \cdots & C_{-M} & \cdots & C_{-2M} \\ C_1 & C_0 & \cdots & C_{-M+1} & \cdots & C_{-2M+1} \\ \vdots & \vdots & & \vdots & \vdots & \vdots & \vdots \\ C_M & C_{M-1} & \cdots & C_0 & \cdots & C_{-M} \\ \vdots & \vdots & & \vdots & \vdots & \vdots & \vdots \\ C_{2M} & C_{2M-1} & \cdots & C_M & \cdots & C_M \end{bmatrix} \begin{bmatrix} b_{-M} \\ b_{-M} \\ \vdots \\ b_0 \\ \vdots \\ b_{-M} \end{bmatrix} = \begin{bmatrix} 0 \\ 0 \\ \vdots \\ 1 \\ \vdots \\ 0 \end{bmatrix} \tag{7.7}$$

it is clear the b_m's can be solved for by solution of a *Toeplitz*[1] set of equations.

Inverse transforming (7.5) gives the time domain restoration formula

$$f(t) = [g_c(t)\Theta_M(\frac{t}{T})] * 2B\mathrm{sinc}(2Bt) \tag{7.8}$$

where $\Theta_M(t)$ is the trigonometric polynomial

$$\Theta_M(t) = \sum_{m=-M}^{M} b_m \, e^{j2\pi mt} \tag{7.9}$$

---

[1] The nm^{th} element of the matrix is a function only of $n - m$.

Note, however, that since

$$g_c(t) = g_c(t) r_\alpha(\frac{t}{T})$$

we only require knowledge of $\Theta_M(t)$ where r_α is unity. Thus we define the periodic function

$$\psi_M(t) = \Theta_M(t) r_\alpha(t). \tag{7.10}$$

Expanding in a Fourier series gives

$$\psi_M(t) = \sum_{n=-\infty}^{\infty} d_n \, e^{j2\pi nt}.$$

The coefficients are

$$
\begin{aligned}
d_n &= \int_{-\frac{1}{2}}^{\frac{1}{2}} \psi_M(t) \, e^{-j2\pi nt} dt \\
&= \int_{-\frac{\alpha}{2}}^{\frac{\alpha}{2}} \Theta_M(t) \, e^{-j2\pi nt} dt \\
&= \alpha \sum_{m=-M}^{M} b_m \, \text{sinc}[\alpha(n-m)]
\end{aligned}
$$

where, in the last step we have used (7.9). From (7.6), we conclude that

$$
d_n = \begin{cases}
\alpha \displaystyle\sum_{m=-M}^{M} b_m \, \text{sinc}[\alpha(n-m)] & ; |\,n\,| \geq M \\[2ex]
\delta[n] & ; |\,n\,| \leq M.
\end{cases}
$$

Note that the d_n's are also the weights of the remaining spectra after restoration. Plots of $\psi_M(t)$ for $\alpha = 0.5$ are shown in Fig. 7.6. Plots of $\psi_2(t)$ for various duty cycles are shown in Fig. 7.7.

In lieu of (7.8), the restoration algorithm pictured in Fig. 7.8 now becomes

$$f(t) = [g_c(t)\psi_M(\frac{t}{T})] * 2B\text{sinc}(2Bt). \tag{7.11}$$

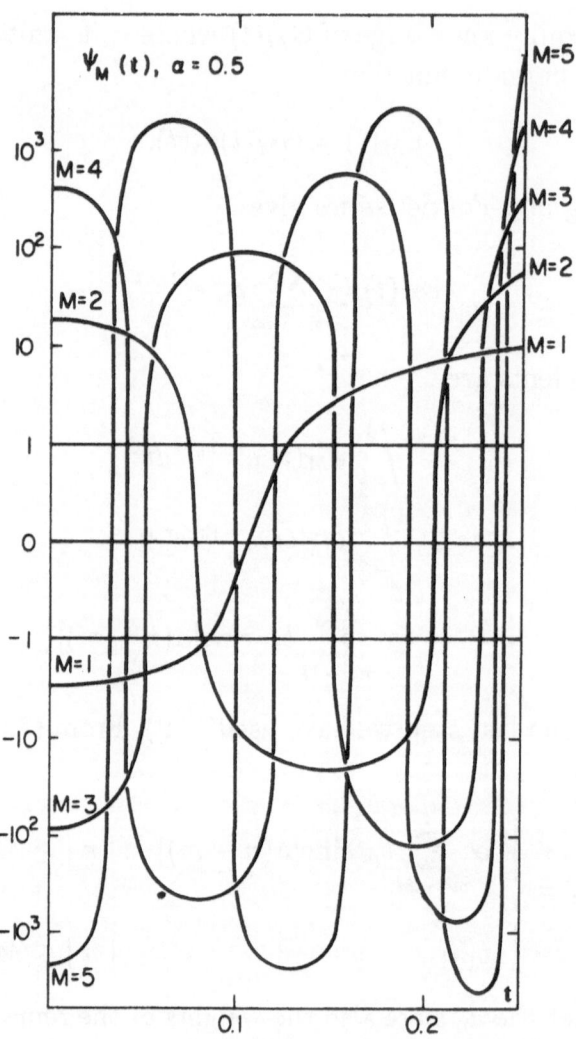

Figure 7.6: Plots of $\psi_M(t) = \psi_M(-t)$ for $\alpha = 0.5$ and $M = 1, 2, 3, 4,$ and 5. The vertical scale is linear for $|\psi_M(t)| < 1$ and logarithmic otherwise.

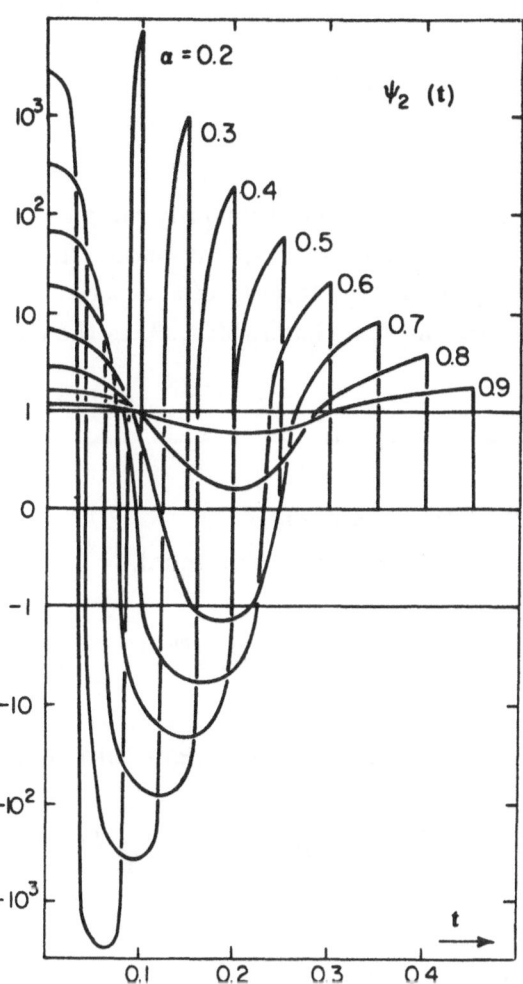

Figure 7.7: Plots of $\psi_2(t)$ for various α. The vertical scale is linear for $\mid \psi_2(t) \mid < 1$ and logarithmic otherwise.

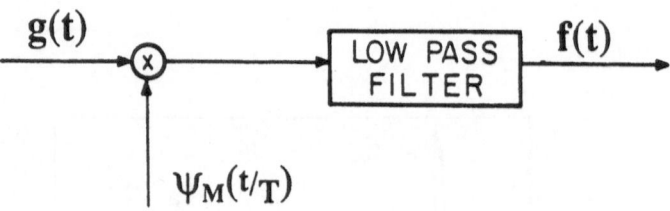

Figure 7.8: Restoration of continuously sampled signals. The periodic function $\psi_M(t)$ is parametrized by the duty
cycle and the severity of aliasing. The low-pass filter
has the same bandwidth, B, as the restored signal.

This is our desired result.

Trigonometric Polynomials
If $f(t)$ is the trigonometric polynomial,

$$f(t) = \sum_{n=-N}^{N} \beta_n \, e^{j2\pi nt/T}$$

then , for $|t| < T/2$, the restoration algorithm in (7.11) becomes

$$f(t) = [\, f(t) \, \Pi(\frac{t}{\alpha T})\psi_M(\frac{t}{T})\,] \, * \, \frac{1}{T} \, h(\frac{t}{T}) \qquad (7.12)$$

where

$$h(t) = \frac{\sin[\pi(2N+1)t]}{\sin(\pi t)} \qquad \longleftrightarrow \qquad H(u) = \sum_{n=-N}^{N} \delta(u-n).$$

This restoration process is illustrated in Fig. 7.9. Note that
$H(uT)$ acts as a sampler in the frequency domain. Plots of $h(t)$
are shown in Fig. 7.10 for various N.

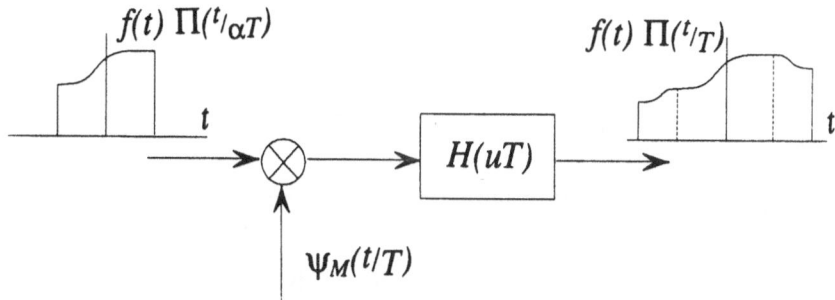

Figure 7.9: Restoration of a period of a trigonometric polynomial known only over the subperiod of $|t| < \frac{1}{2}$.

Proof: We can write (7.11) as

$$f(t) = 2B \sum_{n=-\infty}^{\infty} \int_{(n-\frac{\alpha}{2})T}^{(n+\frac{\alpha}{2})T} g_c(\tau)\, \psi_M(\frac{\tau}{T})\text{sinc}[2B(t-\tau)]d\tau.$$

Since both g_c and ψ_M are periodic, setting $\xi = \tau - nT$ gives

$$f(t) = \int_{-\alpha T/2}^{\alpha T/2} g_c(\xi)\, \psi_M(\frac{\xi}{T})\, k(t-\xi)\, d\xi$$

where

$$k(t) = 2B \sum_{n=-\infty}^{\infty} \text{sinc}[2B(t-nT)].$$

Recognizing that $B = N/T$, we can evaluate this sum in the same manner we evaluated (4.75). The result is

$$k(t) = \frac{\sin[\pi(2N+1)\frac{t}{T}]}{T\sin(\frac{\pi t}{T})}$$

and (7.12) results. Note that we must have the strict inequality $M > 2N$ since, at $M = 2N$, we have first order aliasing due to the Dirac delta nature of the spectrum of $f(t)$.

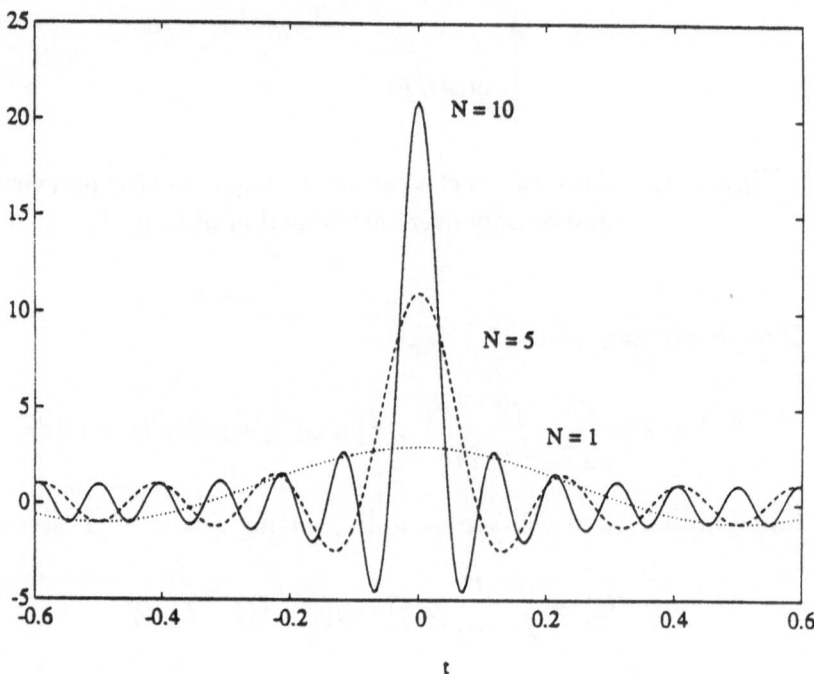

Figure 7.10: Plots of the convolution kernels $h(t)$. When a
trigonometric polynomial of period T is known for
$|t| < \frac{T}{2}$, the convolution of this function with the
known portion of the polynomial (after weighting
with $\psi_M(t)$) results in a whole period of the trigonometric polynomial.

7.1.2   Noise Sensitivity

In this section, we explore the performance of the restoration algorithm in (7.12) in the presence of additive wide sense stationary zero mean noise, $\xi(t)$ [Marks & Kaplan]. Because of linearity, an input of $g_c(t) + \xi(t)$ into the restoration algorithm will yield an output of $f(t) + \eta(t)$, where $\eta(t)$ is the algorithm response to $\xi(t)$ alone. Using (7.11), it follows that

$$\eta(t) = [\xi(t)\,\psi_M(\frac{t}{T})] * 2B\mathrm{sinc}(2Bt).$$

The interpolation noise, $\eta(t)$, is also zero mean although it is not stationary.

The restoration noise level follows as

$$
\begin{aligned}
\overline{\eta^2(t)} &= E\,[\eta^2(t)] \\
&= E\,[\int_{\tau=-\infty}^{\infty} \xi(\tau)\,\psi_M(\frac{\tau}{T})\,2B\mathrm{sinc}\{2B(t-\tau)\}\,d\tau \\
&\qquad \times \int_{\lambda=-\infty}^{\infty} \xi(\lambda)\,\psi_M(\frac{\lambda}{T})\,2B\mathrm{sinc}\{2B(t-\lambda)\}\,d\lambda\,]
\end{aligned}
$$

$$= 4B^2 \int_{-\infty}^{\infty}\int_{-\infty}^{\infty} R_\xi(\tau-\lambda)\psi_M(\frac{\tau}{T})\psi_M(\frac{\lambda}{T})\mathrm{sinc}[2B(t-\tau)]\mathrm{sinc}[2B(t-\lambda)]d\tau d\lambda.$$

By straightforward integral manipulations we have

$$\overline{\eta^2(t)} = \int_{-\infty}^{\infty} R_\xi(\gamma)h(t;\gamma)d\gamma \qquad (7.13)$$

where

$$h(t;\gamma) = 2B\,\psi_M(\frac{\gamma}{T})\mathrm{sinc}[2B(t-\gamma)] \star 2B\,\psi_M(\frac{\gamma}{T})\mathrm{sinc}[2B(t-\gamma)]. \qquad (7.14)$$

The $\star$ denotes autocorrelation with respect to γ. The output noise level in (7.13) is an even periodic function with period T.

7.1.2.1   White Noise

For continuous white noise

$$R_\xi(\tau) = \overline{\xi^2}\delta(\tau). \qquad (7.15)$$

Equation (7.13) becomes

$$\frac{\overline{\eta^2(t)}}{\overline{\xi^2}} = h(t;0)$$

$$= (2B)^2 \text{sinc}^2(2Bt) * \psi_M^2(\frac{t}{T}). \qquad (7.16)$$

From (7.10)

$$\psi_M^2(t) = r_\alpha(t)\Theta_M^2(t)$$

where, from (7.9)

$$\Theta_M^2(t) = \sum_{k=-M}^{M} \sum_{r=-M}^{M} b_k b_r \; e^{j2\pi(k+r)t}. \qquad (7.17)$$

Fourier transforming both sides of (7.16) gives

$$\frac{\overline{\eta^2(t)}}{\overline{\xi^2}} \leftrightarrow 2B\Lambda(\frac{u}{2B})\mathcal{F}\,\psi_M^2(\frac{t}{T})$$

$$= 2B\Lambda(\frac{u}{2B})[TR_\alpha(Tu) * \mathcal{F}\Theta_M^2(\frac{t}{T})] \qquad (7.18)$$

where $\mathcal{F}$ denotes the Fourier transform operator. Substituting (7.4) and the transform of (7.17) into (7.18) gives

$$\frac{\overline{\eta^2(t)}}{\overline{\xi^2}} = 2B\Lambda(\frac{u}{2B}) \sum_{p=-\infty}^{\infty} c_p \sum_{k=-M}^{M} b_k \sum_{r=-M}^{M} b_r \delta[u + \frac{(k+r-p)}{T}]$$

$$= 2B\Lambda(\frac{u}{2B}) \sum_{k=-M}^{M} b_k \sum_{r=-M}^{M} b_r \sum_{q=-M}^{M} c_{k+r-q}\delta(u - \frac{q}{T})$$

where $q = k+r-p$ and we have recognized that the finite extent of the triangle function lets through only $2M + 1$ of the Dirac delta functions. Evaluating $\Lambda(u/2B)$ at $u = q/T$ and inverse transforming gives the desired result:

$$\frac{\overline{\eta^2(t)}}{\overline{\xi^2}} = \sum_{k=-M}^{M} b_k \sum_{r=-M}^{M} b_r \sum_{q=-M}^{M} (2B - \frac{|q|}{T})c_{k+r-q} \; e^{-j2\pi qt/T}.$$

$$(7.19)$$

An illustration of the restoration noise level for various duty cycles for first-degree aliasing is shown in Fig. 7.12. The effects of variation of the aliasing order are illustrated in Fig. 7.11.

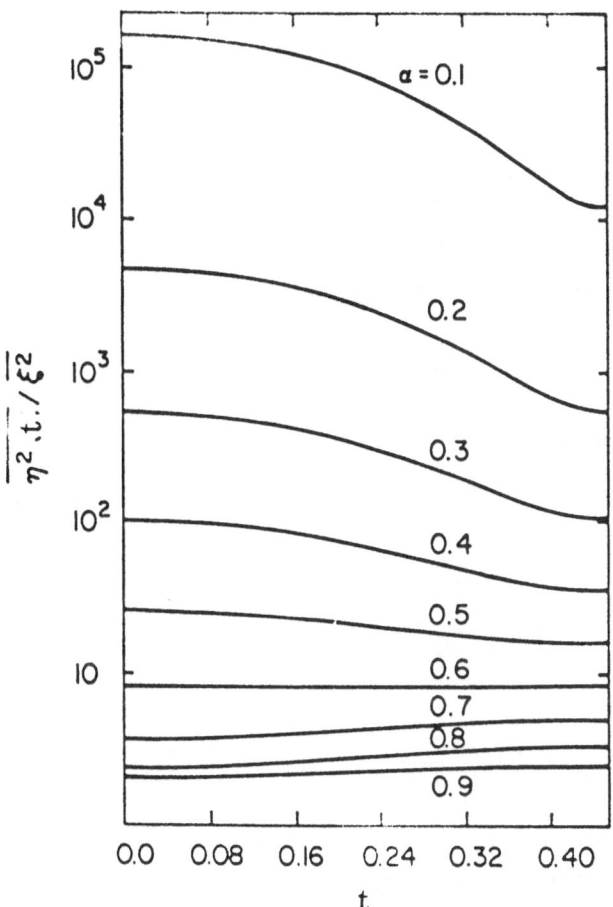

Figure 7.11: NINV for additive white noise for various orders of aliasing M. The values of T corresponding to $M = 1, 2, 3$ are $T = 0.9, 1.4, 1.9$, respectively. Because of symmetry, plots are needed only for $0 < x < \frac{T}{2}$. ($2B = 2$ and $\alpha = 0.6$)

7.1.2.2   Colored Noise

With the aim of placing (7.14) in more tractable form for colored noise, we Fourier tranform with respect to γ using the autocorrelation theorem of Fourier analysis

$$
\begin{aligned}
H(t;\nu) &= \int_{-\infty}^{\infty} h(t;\gamma)\, e^{-j2\pi\nu\gamma} d\gamma \\
&= \mid T\,\Psi_M(T\nu) * [\, e^{-j2\pi\nu t}\, \Pi(\frac{\nu}{2B})]\mid^2 \qquad (7.20)
\end{aligned}
$$

where $\Psi_M(\nu)$ is the Fourier transform of $\psi_M(\gamma)$ and convolution is with respect to ν. Clearly

$$
\Psi_M(\nu) = \sum_{n=-\infty}^{\infty} d_n \delta(\nu + n).
$$

Thus (7.20) becomes

$$
\begin{aligned}
H(t;\nu) &= \mid \sum_{n=-\infty}^{\infty} d_n \exp[-j2\pi(\nu + \frac{n}{T})t]\,\Pi(\frac{\nu + n/T}{2B})\mid^2 \\
&= \mid \sum_{n=-\infty}^{\infty} d_n \exp(\frac{-j2\pi nt}{T})\,\Pi(\frac{\nu + n/T}{2B})\mid^2. \qquad (7.21)
\end{aligned}
$$

Since

$$
\Pi(\frac{\nu + \frac{n}{T}}{2B})\,\Pi(\frac{\nu + \frac{m}{T}}{2B}) = \Pi(\frac{m-n}{4BT})\,\Pi(\frac{\nu + \frac{m+n}{2T}}{2B - \frac{|n-m|}{T}}),
$$

substituting into (7.21) and further recognizing from (7.6) that $d_m = \delta[m]$ for $\mid m \mid < M$ gives

$$
\begin{aligned}
H(t;\nu) &= \Pi(\frac{\nu}{2B}) \\
&\quad + \sum_{n=-M}^{M} \sum_{\substack{\mid m\mid > M \\ \mid n-m\mid \le M}} d_n d_m \Pi(\frac{\nu + \frac{m+n}{2T}}{2B - \frac{|m-n|}{T}})\exp\{\frac{-j2\pi(n-m)t}{T}\} \\
&= \Pi(\frac{\nu}{2B}) \\
&\quad + \sum_{n=-M}^{M} \sum_{\substack{\mid m\mid \le M \\ \mid n-m\mid > M}} d_n d_{n-m}\Pi(\frac{\nu + \frac{2n-m}{2T}}{2B - \frac{|m|}{T}})\exp\{\frac{-j2\pi mt}{T}\}. \qquad (7.22)
\end{aligned}
$$

Using the power theorem, (7.13) can be written as

$$\overline{\eta^2(t)} = \int_{-\infty}^{\infty} S_\xi(u)H(t;u)du \qquad (7.23)$$

where the power spectral density, $S_\xi(u)$, is the Fourier transform of $R_\xi(t)$. Define the odd indefinite integral

$$I_\xi(u;t) = \int_0^u S_\xi(\nu)H(t;\nu)d\nu.$$

Then substituting (7.22) into (7.23) and recognizing that $\eta^2(t)$ is real gives the Fourier series

$$\overline{\eta^2(t)} = 2I_\xi(B;t)$$
$$+ \sum_{n=-M}^{M} \sum_{\substack{|m| \le M \\ |n-m| > M}} d_n d_{n-m} \cos(\tfrac{2\pi mt}{T}) \times$$
$$[I_\xi(\tfrac{m-|m|-2n}{2T} + B;t) - I_\xi(\tfrac{m+|m|-2n}{2T} - B;t)]. \qquad (7.24)$$

For white noise, as in (7.15), $I_\xi(u;t) = \overline{\xi^2}$. The equivalent result in (7.19), however, is in closed form.

For an example application of (7.24), consider the Laplace autocorrelation

$$R_\xi(\tau) = \overline{\xi^2}\,e^{-\lambda|\tau|}$$

where λ is a specified positive parameter. Then

$$I_\xi(u;t) = \overline{\xi^2}\,\frac{\arctan(2\pi u/\lambda)}{\pi}.$$

In the numerical examples to follow, B is set to unity. Figure 7.13 shows the dependence of output noise level on the duty cycle α for first-order aliasing. The dependence of the Laplace parameter is shown in Fig. 7.14 for a fixed duty cycle. As λ increases, adjacent points of the input noise become less correlated and the interpolation noise level decreases. Dependence of the output noise level on order of aliasing is illustrated in Fig. 7.15.

7.1.3   Observations

7.1.3.1   Comparison with the NINV of the Cardinal Series

For certain combinations of the parameters $T, 2B$, and α, the continuously sampled signal can be discretely sampled uniformly at or in excess of the Nyquist rate. Let this rate be denoted by $2W > 2B$. The result is the same as if we had discretely sampled the original signal at a rate of $2W$.

Ex	T	$2B$	α	$2W$	min $\overline{\eta^2(t)}/\overline{\xi^2}$	max $\overline{\eta^2(t)}/\overline{\xi^2}$	$\overline{\eta_0^2}/\overline{\xi^2}$
(a)	1	1.5	0.90	5	0.7460	0.7478	0.7647
(b)	1	1.5	0.70	3	0.7768	0.7842	0.8020
(c)	5	0.3	0.96	1	0.2810	0.2822	0.3754

Table 7.1: Comparison of noise levels for some cases in which the signal can be restored using either the continuously sampled signal-restoration algorithm or the conventional sampling theorem (followed by filtering). The former, in each case, has a lower level. In each case, the Laplace parameter is $\lambda = 2$.

Let $T, 2B$, and α be such that this uniform sampling can be performed. Assume that, as in the previous section, each sample point is perturbed by additive Laplace autocorrelation noise with parameter α. When the noisy samples are interpolated and passed through a filter unity on $\mid u \mid < B$ and zero otherwise, the resulting NINV is given by (5.15). One would expect that the periodic continuous sample restoration would yield a lower noise level since more data are used in the recovery. As the results in Table 7.1 indicate, this is indeed the case.

7.1.3.2   In the Limit as an Extrapolation Algorithm

Keeping αT constant and letting T tend to ∞ alters our algorithm to an extrapolation algorithm if $f(t) \to 0$ as $t \to \infty$. As a consequence, the degree of aliasing, M, becomes unbounded. It is clear from Figures 7.13 and 7.14 that the noise sensitivity in this case increases enormously. This is our first observed indication that the extrapolation problem is ill-posed.

7.1.4   Application to Interval Interpolation

Use of the periodic continuous sample restoration algorithm for interval interpolation is shown in Fig. 7.16. The known data, shown in the top figure, consists of the signal's tails. By selectively throwing away portions of the known data, we can form the continuously sampled signal shown. Our algorithm can be applied and the signal restored.

The unknown interval of length $(1-\alpha)T$ must stay fixed. Note, however, that we have freedom in our choice of T. If we choose T to be small, then we have a small duty cycle and, as is illustrated in Fig. 7.12, a correspondingly large restoration noise level. If we choose T to be large, then the order of aliasing increases and, as witnessed by Figs. 7.11 and 7.14, the restoration noise level is also large. These observations suggest that there might exist some intermediate value of T that has optimal restoration noise properties [Marks & Tseng].

An example where this is the case is pictured in Fig. 7.17, where the normalized interpolation noise variance (NINV) at the origin is plotted versus T for the additive white noise in (2.30). As T increases, the restoration noise level decreases until T is sufficiently large to increase the order of aliasing. Then, as shown, the noise level makes a quantum leap and begins decreasing again until the next order of aliasing is reached. (Values of M are given at the top of the plot.) Note that in this case, the relative minima increase with T and, for minimum restoration noise level, the best choice for T is $1 - \epsilon$ where $0 < \epsilon \ll 1$.

As is shown in Fig. 7.18, the relative minima can also increase with T. Here, the noise has a Laplace autocorrelation with parameter $\lambda = 2$. All other parameters are the same. Increasing λ to 10 again yields decreasing minima as shown in Fig. 7.19. Note that, in any case, the interpolation noise level is finite. By this measure, the interpolation problem is thus well posed.

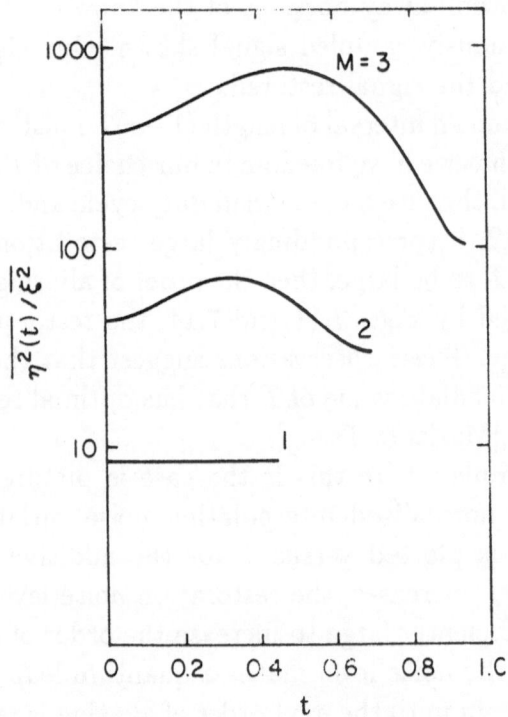

Figure 7.12: NINV for additive white noise for various duty cycles. $2B = 2$ and $T = 0.9$, giving $M =$ first-order aliasing.

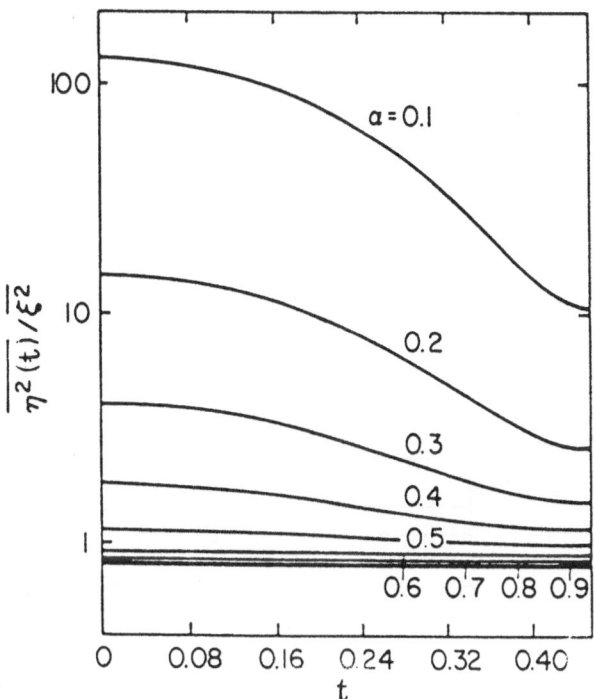

Figure 7.13: NINV for input noise with Laplace autocorrelation
for various duty cycles α. $2B = 2$ and $T = 0.9$,
giving M = first-degree aliasing. The Laplace parameter is $\lambda = 2$.

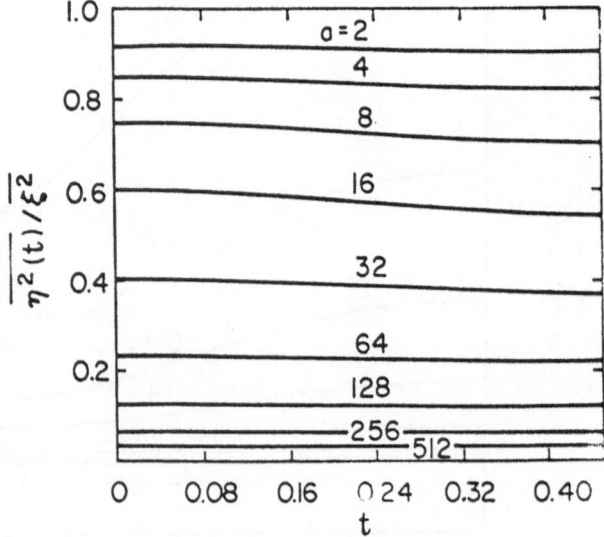

Figure 7.14: NINV for additive input noise with Laplace autocorrelation for various Laplace parameters λ. $2B = 2$ and $T = 0.9$, giving M = first-degree aliasing. The duty cycle is $\alpha = 0.6$.

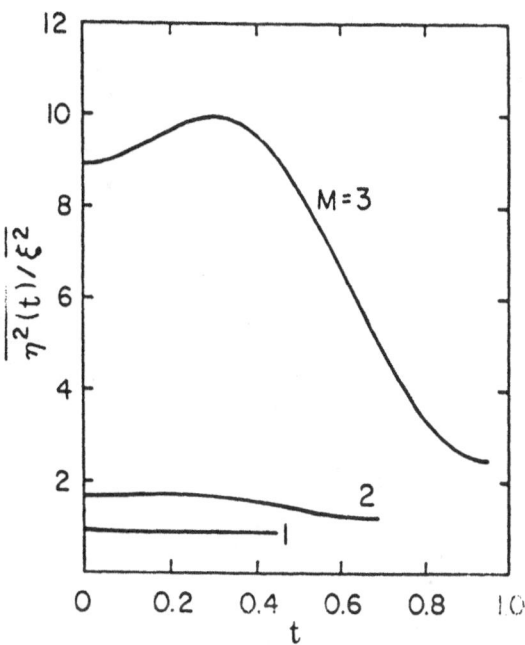

Figure 7.15: NINV for additive input noise with Laplace autocorrelation for various degrees of aliasing. The T values corresponding to $M = 1, 2, 3$ are $T = 0.9, 1.4, 1.9$, respectively. Because of symmetry, plots are needed only for $0 < x < T/2$. ($2B = 2$ and $\alpha = 0.6$, and $\lambda = 2$.)

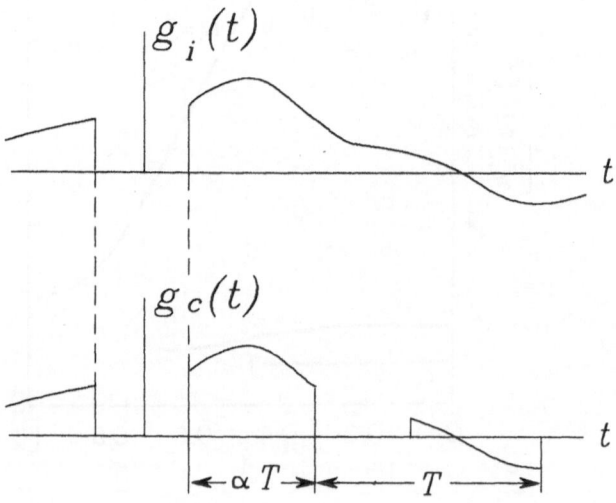

Figure 7.16: Forming a continuously sampled signal (bottom) from the known interpolation problem data (top).

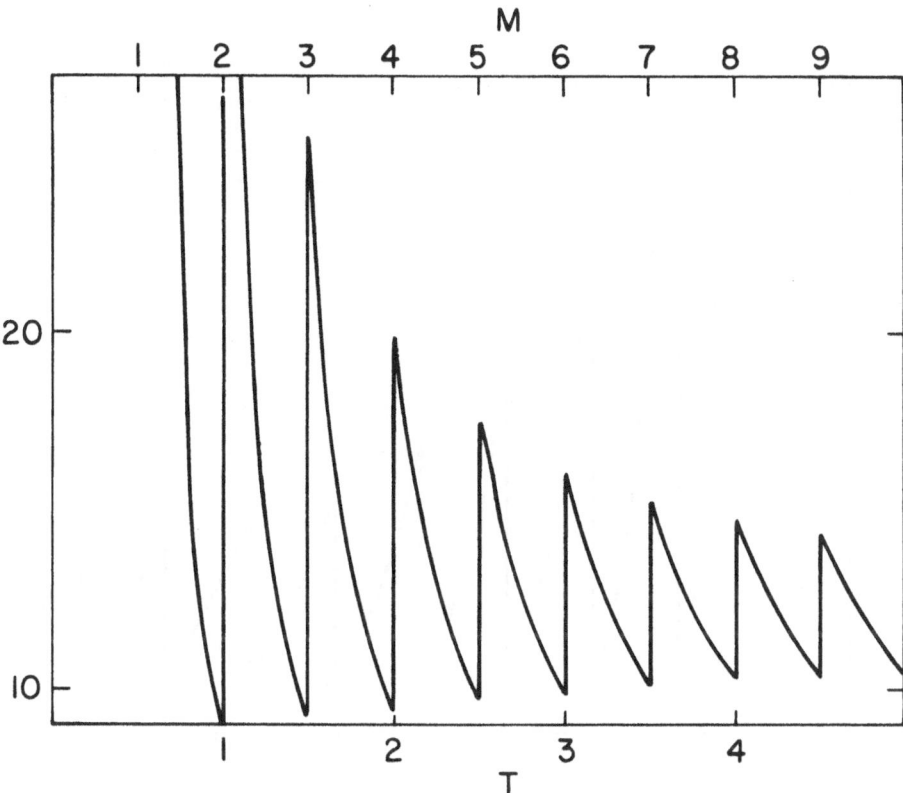

Figure 7.17: Continuous sample restoration of the interval interpolation problem yields this $\overline{\eta^2(0)}/\overline{\xi^2}$ curve when the data are perturbed by white noise. The optimum choice of T is a bit below one $(2B = 2, (1 - \alpha)T = 0.4)$.

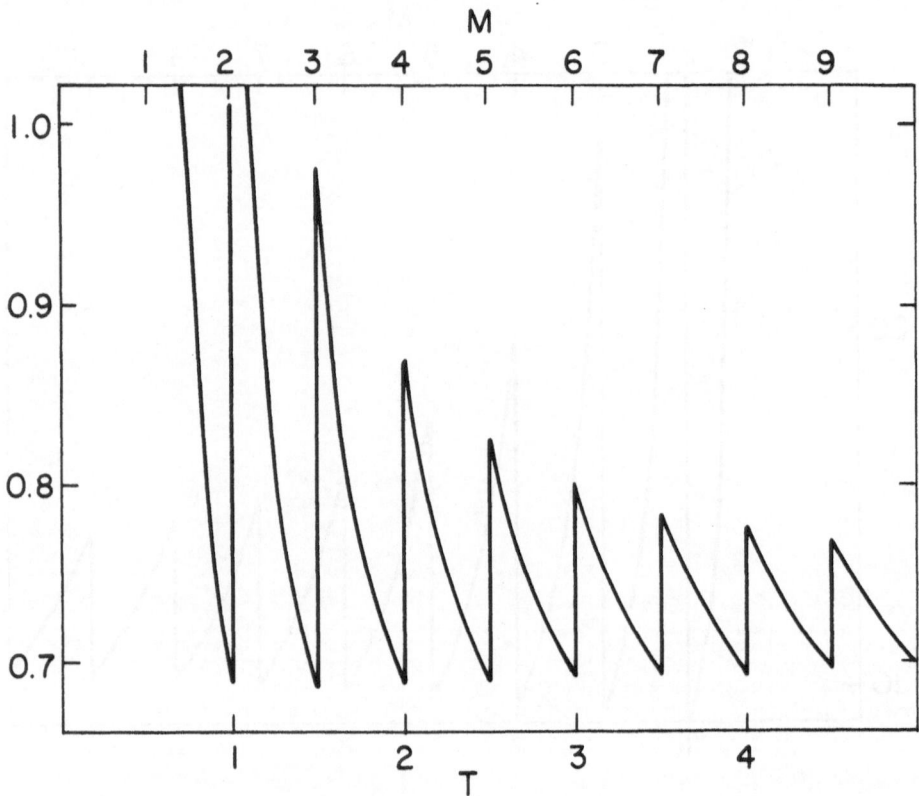

Figure 7.18: Same as Fig. 7.17, except the noise has a Laplace
autocorrelation with parameter $\lambda = 2$. The minima
here increase.

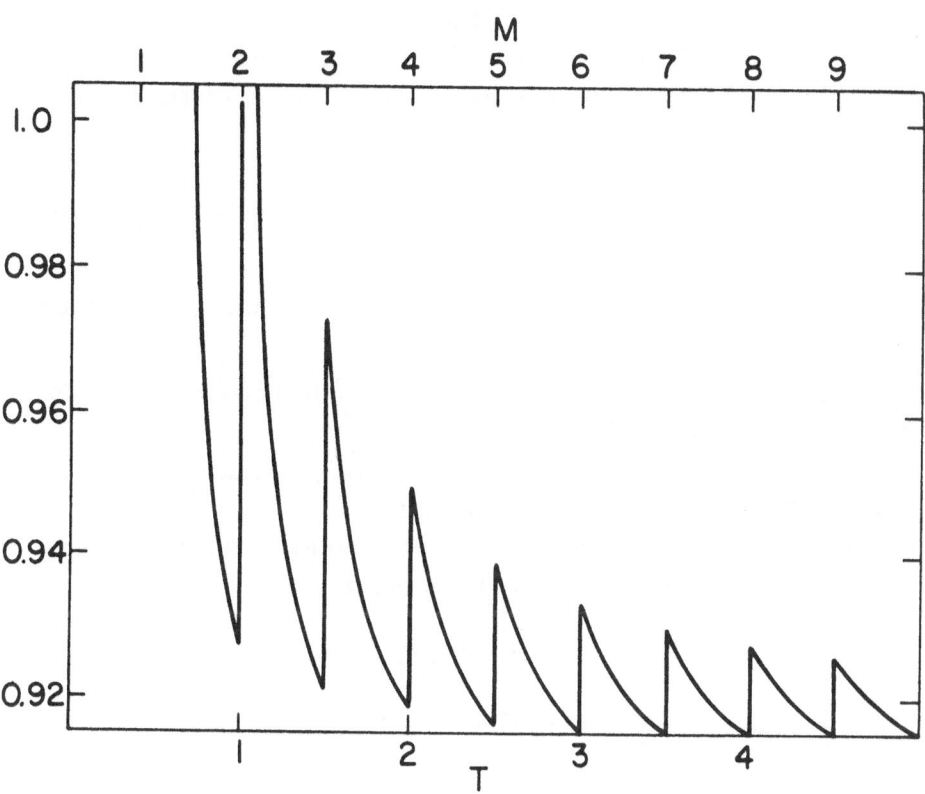

Figure 7.19: Same as Fig. 7.18, except $\lambda = 10$. The minima here decrease. The optimal value of T is a bit below one.

7.2   Prolate Spheroidal Wave Functions

A set of orthogonal functions which prove highly useful in the extrapolation and interval interpolation problems are the *prolate spheroidal wave functions* (PSWF's). Their use in these problems was initially reported by Slepian and Pollak in what today is considered the first in a classic series of papers.

The PSWF's can be defined as the solution of the integral equation:

$$\lambda_n \psi_n(t) = 2B \int_{-T/2}^{T/2} \psi_n(\tau) \, \text{sinc}[2B(t - \tau)] \, d\tau \qquad (7.25)$$

where $0 \leq n < \infty$ and the λ_n's are the eigenvalues. Equivalently

$$\lambda_n \psi_n(t) = [\psi_n(t) \, \Pi(\frac{t}{T})] * 2B\text{sinc}(2Bt). \qquad (7.26)$$

The PSWF's can thus be viewed as the eigenfunctions of low pass filtering signals of finite support.

Although not explicitly stated in the notation, both $\psi_n(t)$ and λ_n are continuous functions of the parameter

$$c = 2BT. \qquad (7.27)$$

Plots of some PSWF's are shown in Fig. 7.20.

7.2.1   Properties

Here we present without proof some significant properties of the PSWF's and their eigenvalues.

(a) The eigenvalues of the PSWF's are real. Note from (7.26), that energy in $\psi_n(t)$ is reduced first by truncation and then by filtering. Thus, each λ_n should have a magnitude less than unity. We will choose them to be positive and will index them in decreasing order:

$$1 > \lambda_0 > \lambda_1 > \cdots > 0. \qquad (7.28)$$

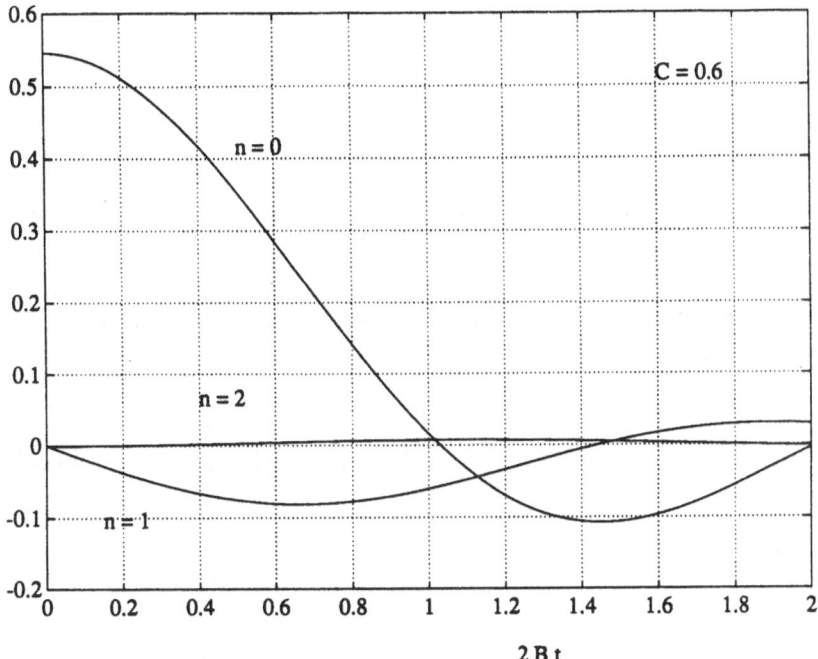

Figure 7.20: $\psi_0(t), \psi_1(t)$, and $\psi_2(t)$ vs. $2Bt$ for $c = 0.6$.

(b) From (7.26), the PSWF's are clearly bandlimited and thus are not affected by low pass filtering:

$$\psi_n(t) = \psi_n(t) * 2B\text{sinc}(2Bt). \tag{7.29}$$

(c) For a given c, the PSWF's are orthonormal on $(-\infty, \infty)$:

$$\int_{-\infty}^{\infty} \psi_n(\tau)\psi_m(\tau)d\tau = \delta[n - m]. \tag{7.30}$$

Furthermore, they form a complete basis set for finite energy bandlimited signals. Thus, if $f(t)$ is bandlimited, then

$$f(t) = \sum_{n=0}^{\infty} a_n\psi_n(t) \tag{7.31}$$

where

$$a_n = \int_{-\infty}^{\infty} f(t)\psi_n(t)dt. \tag{7.32}$$

Like the sampling theorem, convergence of (7.31) is uniform [Gallagher & Wise].

(d) For a given c, the PSWF'S are orthogonal on the interval $|t| < T/2$:

$$\int_{-T/2}^{T/2} \psi_n(\tau)\psi_m(\tau)\, d\tau = \lambda_n\, \delta[n-m]. \qquad (7.33)$$

Furthermore, the PSWF's are complete finite energy functions on the interval $|t| < T/2$. Specifically

$$h(t) = \sum_{n=-N}^{N} b_n \psi_n(t); |t| < \frac{T}{2} \qquad (7.34)$$

where

$$\lambda_n b_n = \int_{-T/2}^{T/2} h(t)\psi_n(t)\, dt.$$

Like the Fourier series, convergence of (7.34) is assured in the mean square sense.

(e) The PSWF's are also eigenfunctions of a sort for the Fourier transform. Specifically

$$\psi_n(t) \leftrightarrow \sqrt{\frac{T}{2B\lambda_n}}\, \psi_n(\frac{Tu}{2B})\quad \Pi(\frac{u}{2B}). \qquad (7.35)$$

Similarly, for the truncated PSWF:

$$\psi_n(t)\, \Pi(\frac{t}{T}) \leftrightarrow \sqrt{\frac{T\lambda_n}{2B}}\quad \psi_n(\frac{Tu}{2B}). \qquad (7.36)$$

This follows from (7.35) and duality.

(f) The PSWF's are difficult to deal with numerically. For our purposes, they will prove to be primarily an analytic tool. Their detail structure, shown in Fig. 7.20, is of secondary interest.

7.2.2   Application to Extrapolation

Let $f(t)$ be bandlimited with known bandwidth B. The extrapolation problem is to regain $f(t)$ from

$$g_e(t) = f(t) \, \Pi(\frac{t}{T}). \tag{7.37}$$

A solution to this problem is obtained by expanding $g(t)$ into a PSWF series:

$$g_e(t) = \sum_{n=0}^{\infty} a_n \psi_n(t) \, \Pi(\frac{t}{T}) \tag{7.38}$$

where

$$\lambda_n a_n = \int_{-T/2}^{T/2} g_e(t) \, \psi_n(t) \, dt. \tag{7.39}$$

Also, since $f(t)$ is bandlimited, it can be expanded as in (7.31). The coefficients of an orthogonal function expansion are unique. Thus, the coefficients in (7.31) and (7.39) are the same. The significant point is that these coefficients can be determined only with knowledge of $g_e(t)$ via (7.39). Then $f(t)$ can be found from (7.31) and our extrapolation is complete.

This result should bother our intuition. For example, telephone conversation waveforms can be considered bandlimited. Our result says that the entirety of a phone conversation can be determined if we know only a word or two in the middle. This, of course, is an unacceptable conclusion.

The resolution of this apparent paradox between mathematics and intuition lies in the fact that our analysis has been to this point deterministic. In practice, the known portion of the signal will be accompanied by some type of noise. To understand how noise affects the algorithm, we must examine the structure of the eigenvalues shown in Fig. 7.21. Fix c. For n below a certain number, the eigenvalues are essentially one. Above that threshold, they are close to zero. A typical plot of λ_n versus n is shown in Fig. 7.22.

Consider, then, the evaluation of the coefficients in (7.39) when either the integral computation and/or $g_e(t)$ is accompanied by a small degree of inexactitude. If n is above the threshold, then division by $\lambda_n \approx 0$ will greatly magnify this error.

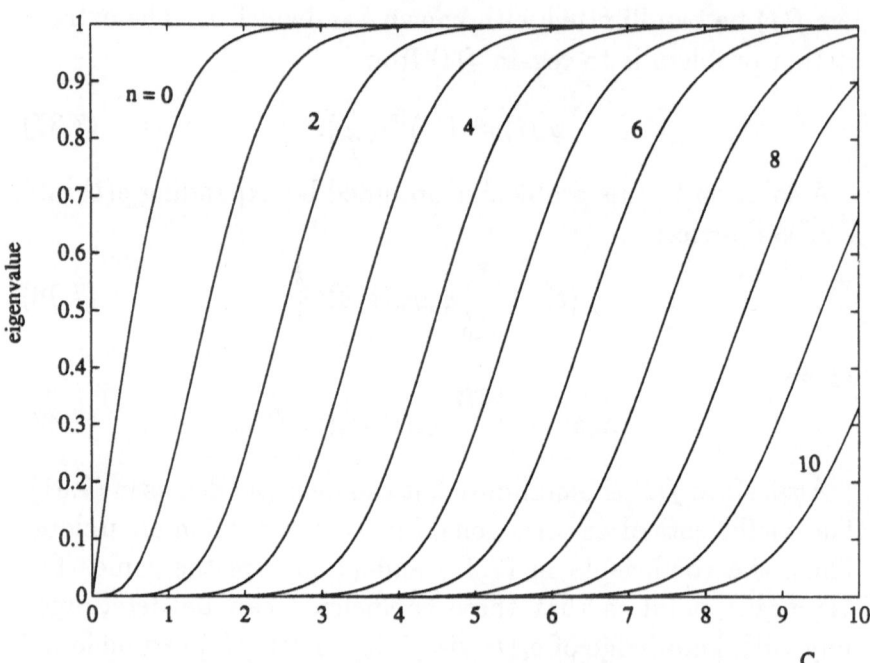

Figure 7.21: Eigenvalues, λ_n, of the PSWF integral equation.

Thus, the a_n coefficients can be only computed reliably up to that threshold which we will call S.

To get a feeling for the value of S, consider again, $g_e(t)$. If we sample this known portion at the Nyquist rate, $2B$, over a time interval of duration T, then the total number of non-zero samples is about $2BT$. This is the time-bandwidth product discussed in Chapter 3. It is also roughly the number of discrete values required to specify $g_e(t)$ to a "good" approximation. We can show empirically that this is the threshold we seek. The value

$$S = 2BT$$

has also been called the *Shannon number* [diFrancia]. Note that $S = c$. Thus, we conclude that in most any practical situation, $g_e(t)$ can be represented by roughly S numbers, be they samples or PSWF coefficients. In very high signal-to-noise ratio situa-

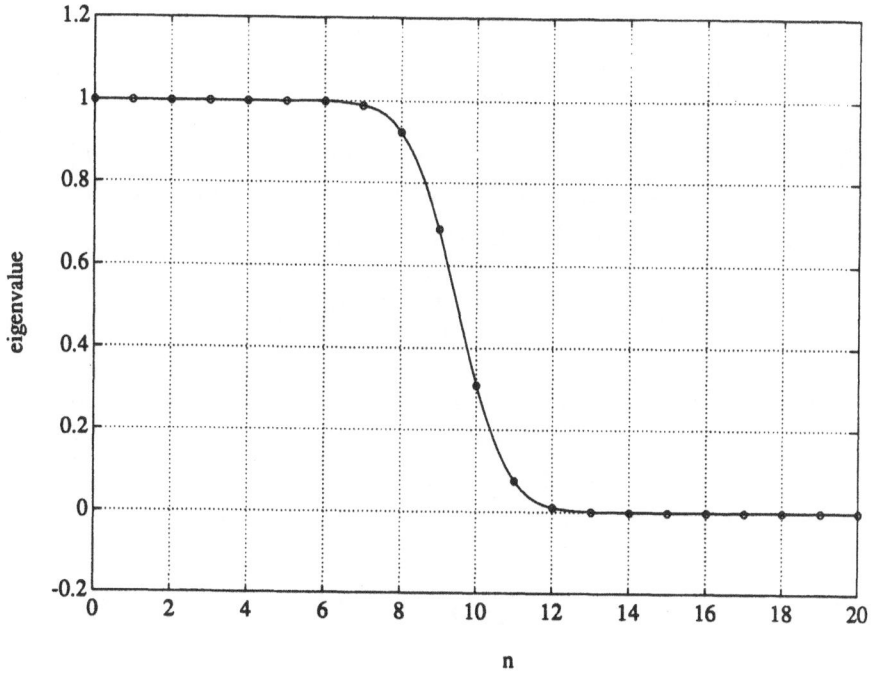

Figure 7.22: Eigenvalues for the PSWF for $c = 10$. The points are connected for clarity.

tions, however, it is possible to add a few degrees of freedom to a truncated signal.

7.2.3 Application to Interval Interpolation

Mathematically, the interpolation problem is similar to extrapolation with a significant difference - interval interpolation is well posed. As is shown in Fig. 7.1c, our given data here is:

$$g_i(t) = f(t)[1 - \Pi(\frac{t}{T})]. \qquad (7.40)$$

From (7.25) and (7.29), we conclude that

$$(1 - \lambda_n)\psi_n(t) = 2B \int_{\tau > \frac{|T|}{2}} \psi_n(\tau)\mathrm{sinc}[2B(t - \tau)]d\tau. \quad (7.41)$$

It follows that the PSWF's are complete for $g_i(t)$ when $f(t)$ is bandlimited. Equation (7.31) applies and the expansion coefficients can be found with knowledge only of $g_i(t)$:

$$a_n = (1 - \lambda_n)^{-1} \int_{\tau > \frac{|T|}{2}} g_i(\tau)\psi_n(\tau)d\tau.$$

Here, we are dividing the integral by a small number for n <u>below</u> S. The difference here is that the number is finite. Thus, finite data error yields a finite amount of interval interpolation error and the problem is well posed.

7.3   The Papoulis-Gerchberg Algorithm

The *Papoulis-Gerchberg algorithm* (PGA) is an ingenious technique for restoring any continuously sampled bandlimited signal without directly using PSWF's. Indeed, the algorithm requires only the operations of filtering and truncation.

The PGA was first discovered by Papoulis (1973-74) but was first published in an archival journal, independently, by Gerchberg. DeSantis and Gori, independently, published the algorithm proof shortly after Gerchberg's paper appeared.

The PGA is applicable to each of the continuous sampling problems illustrated in Fig. 7.1 when $f(t)$ is bandlimited. The algorithm is most easily proved for the cases of extrapolation and interval interpolation. Since most of the work performed has been on the extrapolation problem, this will be our main focus. As before, our results will be ill-posed. Problems - not algorithms - are well or ill posed.

7.3.1   The Basic Algorithm

The PGA is illustrated in Fig. 7.23 for the case of extrapolation. The known portion of the signal is $g_e(t)$ as shown in Fig.

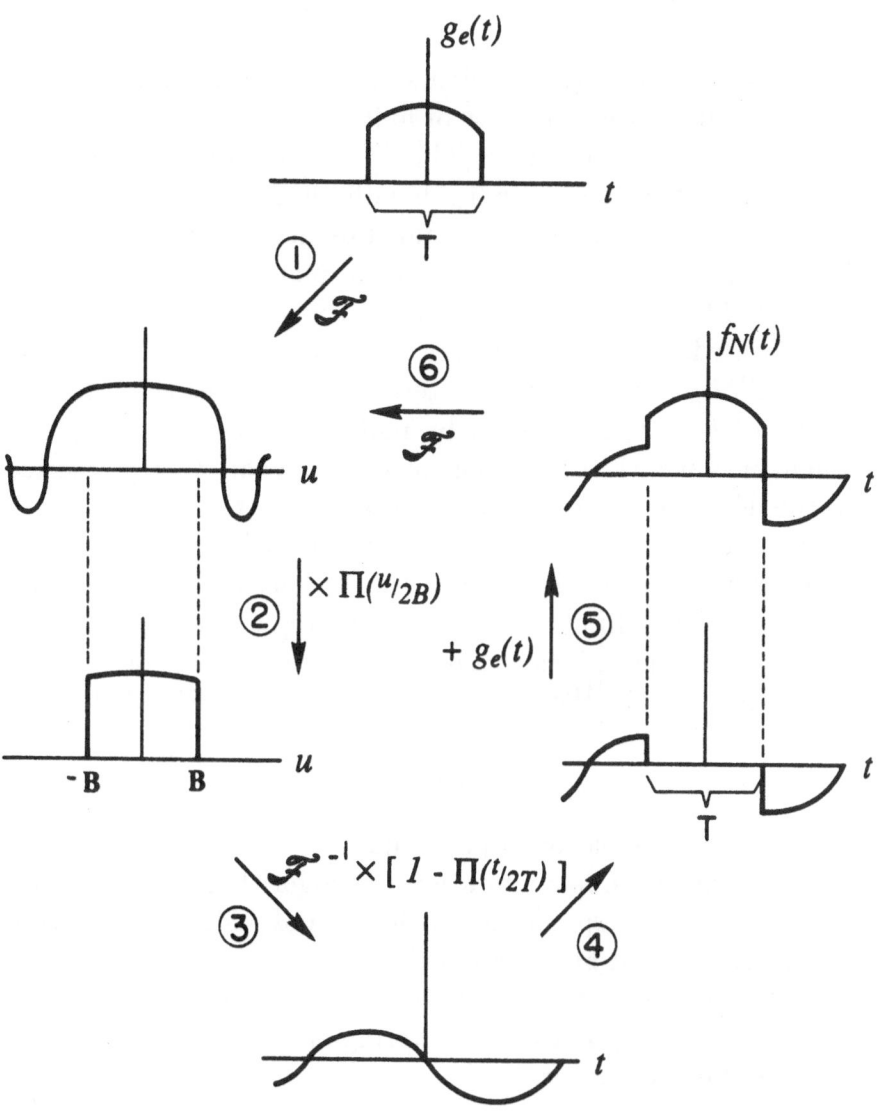

Figure 7.23: Illustration of the PGA applied to extrapolation.

7.1b. We know, secondly, that our signal to be restored, $f(t)$, is bandlimited with bandwidth B. The PGA iterates back and forth between the time and frequency domains reinforcing these criteria.

Beginning with $g_e(t)$, the first step in the algorithm is Fourier transformation. Since $g_e(t)$ is of finite extent, its transform will be identically zero nowhere. This is contrary to our knowledge that the signal to be restored is bandlimited. Thus, we make the spectrum that of a bandlimited function in step 2 by multiplying by $\Pi(u/2B)$. Step 3 is inverse transformation back to the time domain. This signal, shown at the bottom of Fig. 7.23, is clearly bandlimited. It is not, however, equal to $g_e(t)$ on the interval $|t| < T/2$. To impose this criterion, we first set the signal to zero on $|t| < T/2$ in step 4 by multiplying by $1 - \Pi(u/2B)$. Then, in step 5, the known portion of the signal, $g_e(t)$, is inserted in the dead space. We will call this signal $f_1(t)$.

In general, $f_1(t)$ will be a discontinuous function which, in turn, cannot be bandlimited. Thus, we need to reimpose the criterion of bandlimitedness. Thus, in step 6, we begin the set of the same operations again. Denote the results of the Nth iteration by $f_N(t)$. We will show in the next section that

$$\lim_{N \to \infty} f_N(t) = f(t).$$

Thus, our extrapolation is performed.

Note that, by simple alteration, the algorithm can be applied to the interval interpolation, prediction and, indeed, to any continuously sampled bandlimited signal.

Let's operationally compress the PGA of Fig. 7.23. Steps 1, 2 and 3 are simply a low pass filtering operation. We define the low pass filter operator

$$\mathbf{B}_B h(t) = h(t) * 2B\,\text{sinc}(2Bt).$$

The operation of discarding the center can be modeled by $1 - \mathbf{D}_T$ where the duration limiting operator is defined by:

$$\mathbf{D}_T h(t) = h(t)\,\Pi(\frac{t}{T}).$$

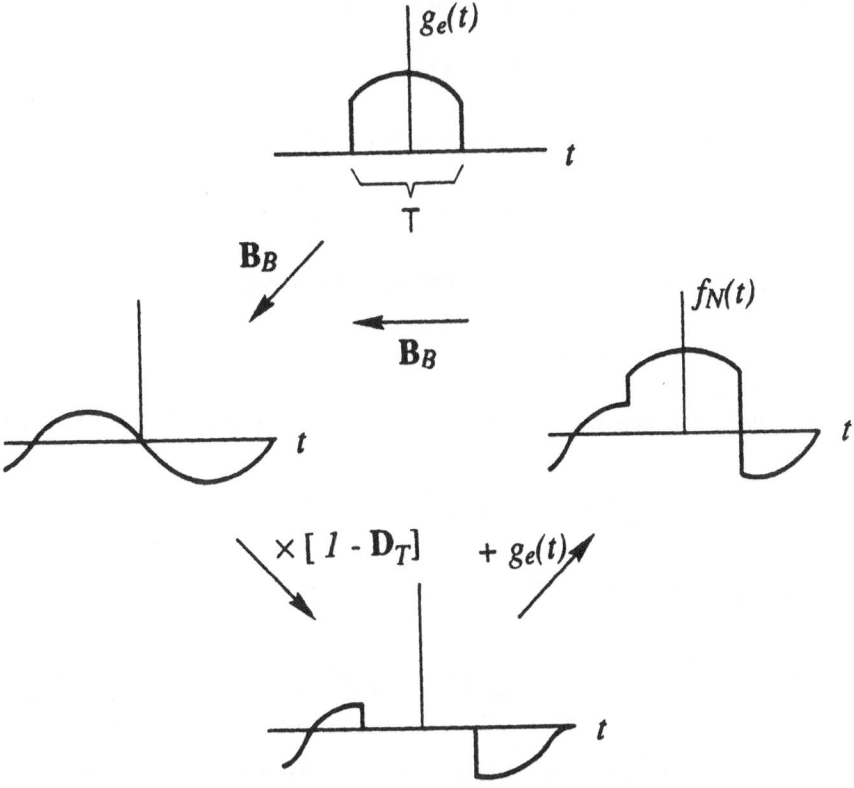

Figure 7.24: An equivalent illustration of the PGA algorithm.

With this notation, the PGA can be written as

$$f_{N+1}(t) = g_e(t) + [1 - \mathbf{D}_T]\mathbf{B}_B f_N(t) \tag{7.42}$$

with initialization

$$f_0(t) = g_e(t). \tag{7.43}$$

The PGA using operators is illustrated in Fig. 7.24.

An alternate derivation of the PGA makes use of the identity

$$\mathbf{B}_B f(t) = f(t).$$

This follows from the bandlimitedness of $f(t)$. Thus

$$g_e(t) = [1 - (1 - \mathbf{D}_T)\mathbf{B}_B]f(t).$$

If the operator in square brackets can be inverted, then we have

$$f(t) = [1 - (1 - \mathbf{D}_T)\mathbf{B}_B]^{-1}g_e(t). \tag{7.44}$$

For inversion, we generalize the geometric series in (5.13) and write:

$$f(t) = \sum_{n=0}^{\infty}[(1 - \mathbf{D}_T)\mathbf{B}_B]^n g_e(t).$$

In the spirit of truncation, define

$$f_N(t) = \sum_{n=0}^{N}[(1 - \mathbf{D}_T)\mathbf{B}_B]^n g_e(t). \tag{7.45}$$

Equations (7.42) and (7.43) follow as a direct consequence.

Again, the PGA can be used for restoration of any bandlimited signal that is continuously sampled in the absence of noise. The algorithm for interval interpolation is, for example:

$$f_{N+1}(t) = g_i(t) + \mathbf{D}_T\mathbf{B}_B f_N(t) \tag{7.46}$$

where

$$\begin{aligned} g_e(t) &= (1 - \mathbf{D}_T)f(t) \\ &= f_0(t). \end{aligned} \tag{7.47}$$

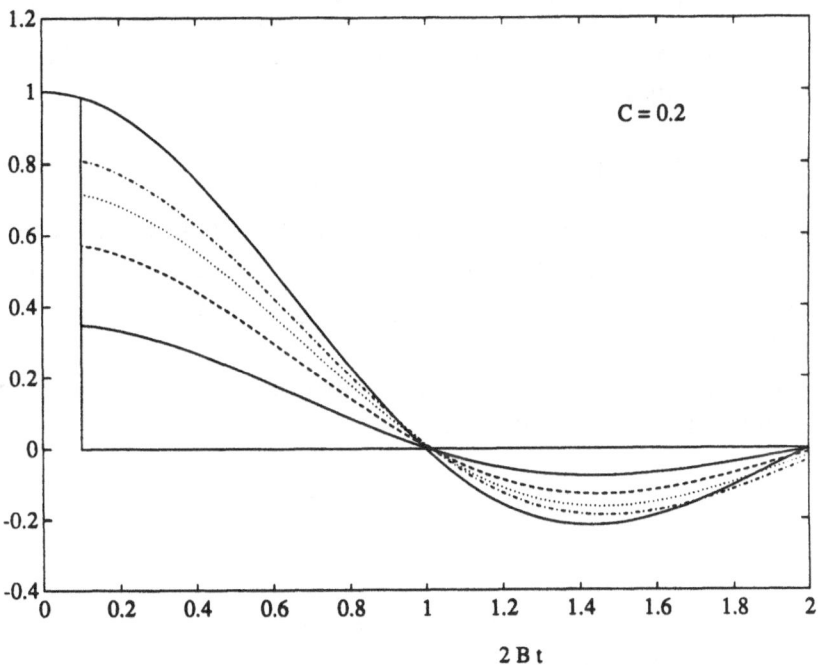

Figure 7.25: Numerical results of the PGA for extrapolation.

The interpolation equivalents of (7.44) and (7.45) are, respectively

$$f(t) = [(1 - \mathbf{D}_T)\mathbf{B}_B]^{-1}g_i(t) \tag{7.48}$$

and

$$f_N(t) = \sum_{n=0}^{N}[\mathbf{D}_T\mathbf{B}_B]^n g_i(t). \tag{7.49}$$

A numerical example of the PGA for extrapolation is shown in Fig. 7.25 for the case of a sinc. "Good" convergence takes place in only eight iterations. We must remark, however, that sincs extrapolate well. Furthermore, the only noise in the unknown signal is computational round off error.

7.3.2   Proof of the PGA using PSWF's

Here, we offer a proof of the PGA using PSWF's for the case of extrapolation [Papoulis (1975), DeSantis and Gori, Marks and Smith]. An alternate proof in a vector space setting, is given by Youla.

We begin by noting that the identities in (7.25) and (7.29) can be respectively written as

$$\mathbf{B}_B\mathbf{D}_T\psi_n(t) = \lambda_n\psi_n(t)$$

and

$$\mathbf{B}_B\psi_n(t) = \psi_n(t).$$

Also, from (7.38) we can write the known portion of the signal as

$$g_e(t) = \sum_{n=0}^{\infty} a_n\mathbf{D}_T\psi_n(t). \tag{7.50}$$

From (7.42) and (7.43), the first iteration of the PGA can be written:

$$\begin{aligned} f_1(t) &= g_e(t) + (1 - \mathbf{D}_T)\mathbf{B}_B + \sum_{n=0}^{\infty} a_n\mathbf{D}_T\psi_n(t) \\ &= g_e(t) + (1 - \mathbf{D}_T)\sum_{n=0}^{\infty} a_n\lambda_n\psi_n(t). \end{aligned} \tag{7.51}$$

Expanding $g_e(t)$ via (7.50) and applying the second iteration gives:

$$f_2(t) = g_e(t) + (1 - \mathbf{D}_T) \sum_{n=0}^{\infty} a_n(2\lambda_n - \lambda_n^2)\psi_n(t).$$

Repeating

$$f_3(t) = g_e(t) + (1 - \mathbf{D}_T) \sum_{n=0}^{\infty} a_n(3\lambda_n - 3\lambda_n^2 + \lambda_n^3)\psi_n(t)$$

$$f_4(t) = g_e(t) + (1 - \mathbf{D}_T) \sum_{n=0}^{\infty} a_n(4\lambda_n - 6\lambda_n^2 + 4\lambda_n^3 - \lambda_n^4)\psi_n(t).$$

This is a sufficient number of iterates to recognize that the coefficients of the eigenvalues are binomial. We can show by induction that

$$f_N(t) = g_e(t) + (1 - \mathbf{D}_T) \sum_{n=0}^{\infty} a_n[1 - (1 - \lambda_n)^N]\psi_n(t). \quad (7.52)$$

Specifically, from (7.51), the result is true for $N = 1$. We will assume (7.52) and show the corresponding equation for $N + 1$ follows. An additional iteration gives

$$g_e(t) \quad + \quad (1 - \mathbf{D}_T)\mathbf{B}_B f_{N+1}(t)$$

$$= \quad g_e(t) + (1 - \mathbf{D}_T) \sum_{n=0}^{\infty} a_n[\lambda_n + (1 - \lambda_n)\{1 - (1 - \lambda_n)\}^N]\psi_n(t).$$

After a small bit of algebra, this relationship becomes (7.52) for $N + 1$. The validity of (7.52) is thus proved.

Since, from (7.28),

$$0 < \lambda_n < 1 \qquad (7.53)$$

we conclude that

$$\lim_{N\to\infty}[1 - (1 - \lambda_n)^N] = 1$$

and, in the limit,

$$
\begin{aligned}
\lim_{N \to \infty} f_N(t) &= g_e(t) + (1 - \mathbf{D}_T) \sum_{n=0}^{\infty} a_n \psi_n(t) \\
&= \Pi(t/T) f(t) + [1 - \Pi(t/T)] \, f(t)
\end{aligned}
$$

where we have used (7.37) and (7.31). This completes our proof of the PGA.

7.3.3   Geometrical Interpretation in a Hilbert Space

An intriguing geometrical interpretation of the PGA is shown in Fig. 7.26 for the case of extrapolation [Youla]. Three subspaces are illustrated. The first, on the horizontal axis, consists of all duration limited signals for which

$$
\mathbf{D}_T h_1(t) = h_1(t).
$$

Thus, as shown, our known signal, $g_e(t)$, lies in this space.

The second subspace corresponds to the vertical axis and represents all signals with a dead space in the middle. That is

$$
(1 - \mathbf{D}_T) h_2(t) = h_2(t).
$$

Note that these two subspaces are orthogonal since

$$
\int_{-\infty}^{\infty} h_1(t) h_2(t) dt = 0.
$$

Thus, they are shown at 90° in Fig. 7.26.

The final subspace is that of all bandlimited signals with bandwidth B. If a signal is in this space,

$$
\mathbf{B}_B h_3(t) = h_3(t).
$$

We now will illustrate the PGA illustrated in Fig. 7.24 in this vector space setting. The signal we wish to retrieve, $f(t)$, lies on the bandlimited signal subspace. We only have knowledge of the projection of this signal onto the duration limited subspace. Referring again to Fig. 7.24, the first step in the PGA is

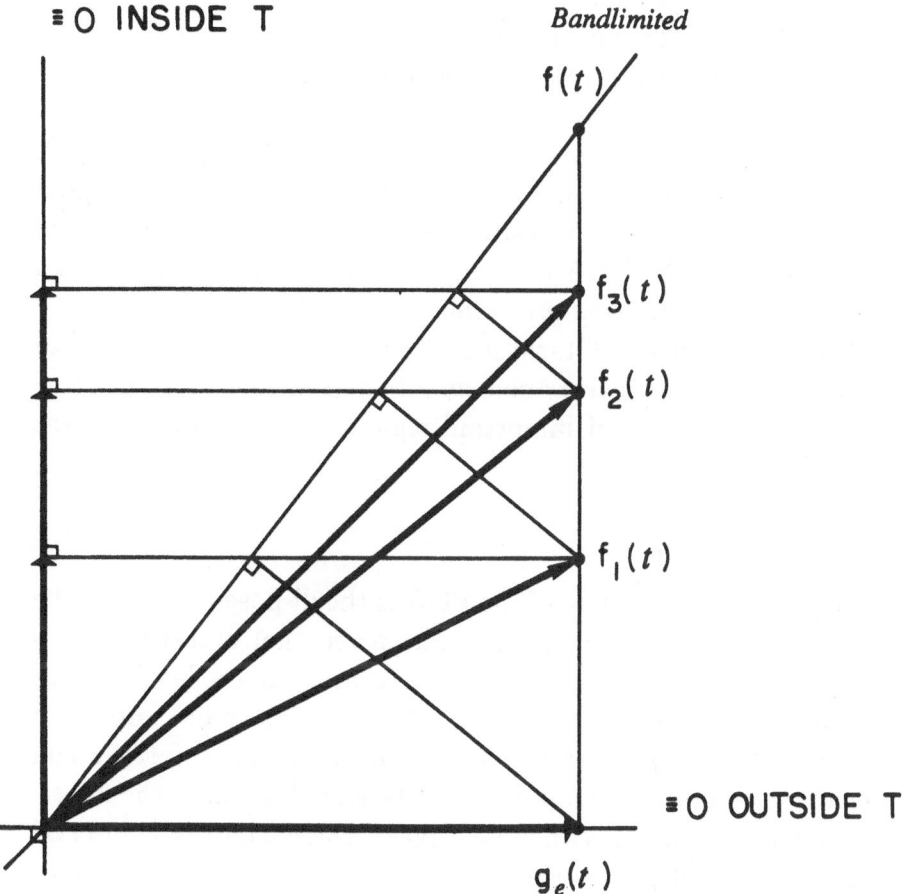

Figure 7.26: Youla's illustration of the convergence of the algorithm in Fig. 7.24 in a Hilbert space.

to bandlimit the signal, $g_e(t)$. This is done in Fig. 7.26 by projecting $g_e(t)$ onto the bandlimited signal subspace. In the next step, we discard the middle of the signal. This is performed by a second projection onto the vertical subspace. Next, we add in the known portion of the signal. This is performed by vectorially adding the projection on the vertical subspace to the known signal, $g_e(t)$. As is shown in Fig. 7.26, the result is $f_1(t)$. Beginning with $f_1(t)$, we repeat the process and generate $f_2(t)$. One can see geometrically that each iteration produces a signal closer and closer to our desired result. Appealing to this signal space format, the PGA can be proven as a special case of *Von Neumann's alternating projection theorem*.

Youla and others [Stark] have generalized these concepts to projection onto arbitrary convex sets and have, as a result, formulated a number of important signal recovery and synthesis algorithms.

7.3.4   Remarks

A fundamental problem of the PGA is the ill-posed nature of the extrapolation problem. We note, however, that extrapolation is ill-posed in the sense that the restoration noise level cannot be bounded. This makes sense. The uncertainty of restoration should in general increase as we remove ourselves farther and farther from the known portion of the signal. Clearly, the uncertainty of restoration with knowledge of a signal only on a interval of length T will be enormous at a distance of, say, $10^{10}T$ away. Thus, extrapolation is ill-posed in a global sort of way. As has been mentioned, however, one would expect "good" results near to where the (smooth) signal is known.

A problem is ill-posed because insufficient information about the restored signal has been provided. To *regularize* such problems, either additional information about the signal must be provided, or the class of allowable solutions restricted.

7.4   Exercises

7.1 The periodic continuously sampled signal at the bottom of Fig. 7.16 can be written:

$$g_c(t) = f(t)\left[1 - r_{1-\alpha}(t/T)\right].$$

Assume $f(t)$ has bandwidth B. Specify how to find the trigonometric polynomial, $\Theta_M(t)$, such that

$$f(t) = [g_c(t)\psi_M(t)] * 2B\operatorname{sinc}(2Bt)$$

where

$$\psi_M(t) = \Theta_M(t)[1 - r_{1-\alpha}(t)].$$

7.2 (a) If $f(t)$ is real, $F(u)$ is *Hermetian*:

$$F(u) = F^*(-u).$$

Thus, knowledge of $F(u)$ for $u > 0$ is sufficient to uniquely specify $f(t)$. Write the formula for $f(t)$ in terms of $F(u)$ for $u > 0$.

(b) Consider, then, first order aliasing for periodic continuous sampling of a bandlimited signal. As shown in Fig. 7.27, we add two weighted versions of $G(u)$ to rid ourselves of the positive frequency overlap. Specify $F(u)$ for positive u in terms of $G(u)$ and $G(u-\frac{1}{T})$.

(c) Find $f(t)$ directly from $g(t)$.

(d) Draw a diagram of your interpolation algorithm akin to Fig. 7.8. All operations should be real.

7.3 Find the PSWF expansion of $2B\operatorname{sinc}[2B(t-\tau)]$.

7.4 The cardinal series can be viewed as an orthogonal expansion using $\{\operatorname{sinc}(2Wt - n) \mid -\infty < n < \infty\}$ as the basis function set

(a) Show these functions form an orthogonal set.

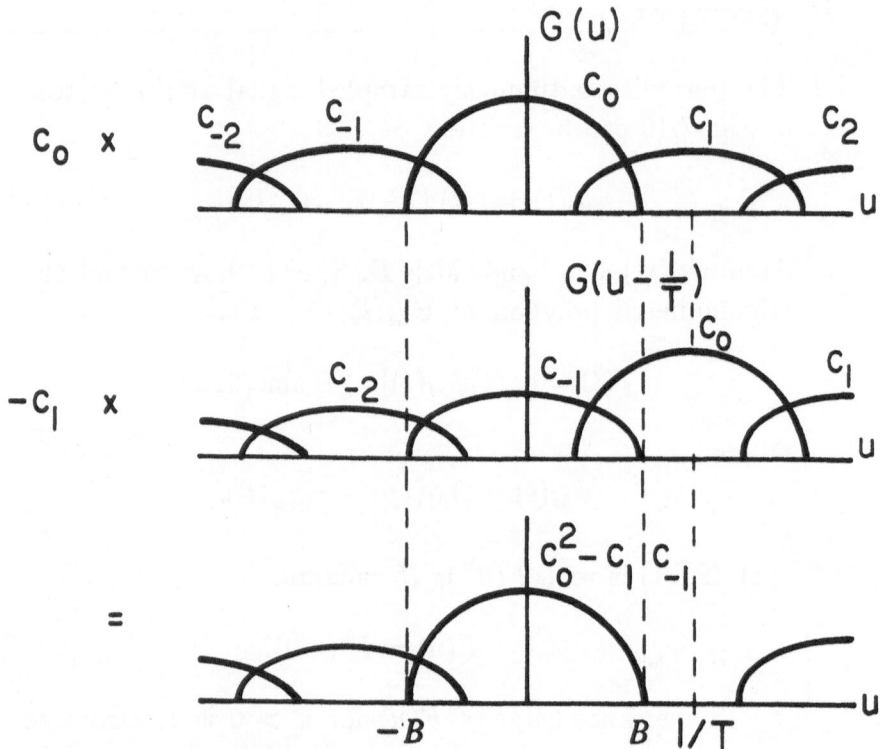

Figure 7.27: Removing the first order spectrum by substracting two weighted and shifted versions of the degraded spectrum.

(b) Express the PSWF expansion coefficients, a_n, in terms of the samples $f(n/2B)$ assuming $f(t)$ is bandlimited. The result is an infinite sum. Is it well-posed?

(c) Similarly, express, $f(t/2B)$ as a weighted sum of PSWF coefficients, a_n. Is it well-posed?

7.5 (a) Show that (7.33) follows as a consequence of (7.30) and (7.25).

(b) Show that (7.33) follows as a consequence of (7.30) and (7.35).

7.6 Clearly, if $\psi_n(t)$ is a solution of (7.25), then so is *const*

$\times \psi_n(t)$. Is there an ambiguity here in our definition of the PSWF? If not, how have we removed it?

7.7 *Parseval's Theorem for PSWF's*

(a) For the expansion in (7.31), show that the energy of $f(t)$ is

$$E = \sum_{n=0}^{\infty} | a_n |^2 .$$

(b) Derive a similar expression for the energy of $h(t)$ in (7.34).

7.8 We pass a signal, $x(t) = x(t) \, \Pi(t/T)$, through a low pass filter with bandwidth B. The output is $y(t)$. Assuming $x(t)$ has unit energy, what input will yield an output with the maximum energy? What is this energy?

7.9 Prove the PGA as applied to interval interpolation.

7.10 For the extrapolation problem, $g_e(t)$ in (7.37) can be expressed in terms of a Fourier series:

$$g_e(t) = \sum_{n=-\infty}^{\infty} g_n \exp(\frac{j2\pi nt}{T}) \, \Pi(\frac{t}{T}).$$

(a) Find the a_n's in (7.39) directly as an infinite weighted sum of the Fourier coefficients, g_n. Is this restoration well-posed or ill-posed?

(b) Conversely, find the g_n's. Comment again on the posedness.

7.11 Show that (7.42) follows as a consequence of (7.45).

7.12 (a) Show that the three subspaces in Fig. 7.26 have only the signal that is identically zero (the origin) as a common element.

(b) Using the same subspaces as are in Fig. 7.26, illustrate geometrically the PGA for interval interpolation.

7.13 (a) Assume we know apriori that the signal to be restored is non-negative, *i.e.*, $f(t) \geq 0$. Incorporate this constraint into the PGA for extrapolation.

(b) Suppose $f(t) = R_\xi(t)$ is an autocorrelation. Incorporate this information into the PGA for extrapolation.

7.14 The geometric series, $\sum_{n=0}^{\infty} z^n = 1/(1 - z)$, converges if $|z| < 1$. Similarly, a sufficient condition for the equivalent equation:

$$(1 - \mathbf{H})^{-1} h(t) = \sum_{n=0}^{\infty} \mathbf{H}^n h(t)$$

is that

$$\| \mathbf{H} \| < 1.$$

The operator norm can be defined by

$$\| \mathbf{H} \| = \sup_{\| h(t) \| = 1} \| \mathbf{H} h(t) \|$$

where the L_2 norm is defined by

$$\| y(t) \|^2 = \int_{-\infty}^{\infty} | y(t) |^2 \, dt$$

and sup denotes 'supremum'.

(a) For the extrapolation algorithm, from (7.44), $\mathbf{H} = (1 - \mathbf{D}_T)\mathbf{B}_B$. Compute $\| \mathbf{H} \|$.

(b) For interval interpolation, from (7.46), $\mathbf{H} = \mathbf{D}_T \mathbf{B}_B$. Compute $\| \mathbf{H} \|$.

7.15 We apply the PGA to the restoration of a single lost sample at the origin of an oversampled signal. The given signal, $g(t)$, is shown in Figure 7.28. It can be written as

$$g(t) = r \sum_{n \neq 0} f(nT)\text{sinc}(2Wt - rn)$$

where $T = 1/2W$ and $f(t)$ has bandwidth of $B < W$. From $g(t)$ we wish to find $f(0)$. As is shown in Fig. 7.28,

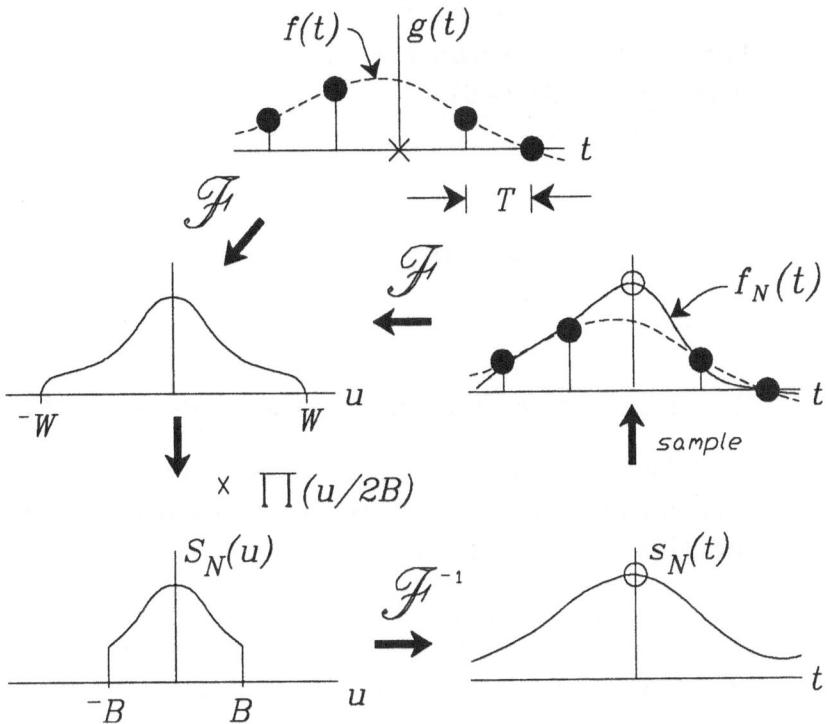

Figure 7.28: Application of the PGA to the restoration of a single lost sample at the origin of an oversampled signal.

we first Fourier transform this sequence and multiply by $\Pi(u/2B)$ to form $S_N(u)$. The inverse Fourier transform of this expression is evaluated at the origin and used as an estimate for the lost sample. After the Nth iteration, we have

$$s_N(t) = T\, f_N(0) + g(t).$$

The iteration is repeated. Evaluate $f_\infty(0)$ and compare your answer with (4.9).

REFERENCES

R.P. Boas, **Entire Functions**, Academic Press, NY, 1954.

A.B. Carlson, **Communications Systems, An Introduction to Signals and Noise in Electrical Communication, 2nd Ed**, McGraw-Hill, New York, 1975.

K.F. Cheung, R.J. Marks II and L.E. Atlas "Convergence of Howard's mimimum negativity constraint extrapolation algorithm", *Journal of the Optical Society of America A*, vol.5, pp.2008-2009 (1988).

P. DeSantis and F. Gori "On an iterative method for super-resolution", *Optica Acta*, vol.22, pp.691-695 (1975).

G.T. diFrancia "Degrees of freedom of an image", *Journal of the Optical Society of America*, vol. 59, pp.799-804 (1969).

N.C. Gallagher Jr. and G.L. Wise "A representation for bandlimited functions", *Proc. IEEE*, vol. 63, p.1624 (1975).

S.J. Howard "Continuation of discrete Fourier spectra using minimum-negativity constraint", *Journal of the Optical Society of America*, vol.71, pp.819-824 (1981).

R.J. Marks II "Restoration of continuously sampled band-limited signals from aliased data", *IEEE Transactions on Acoustics, Speech and Signal Processing*, vol. ASSP-30, pp.937-942 (1982).

R.J. Marks II and D. Kaplan "Stability of an algorithm to restore continuously sampled bandlimited images from aliased data", *Journal of the Optical Society of America*, vol. 73, pp.1518-1522 (1983).

R.J. Marks II and D.K. Smith "Gerchberg-type linear deconvolution and extrapolation algorithms" in **Transformations in Optical Signal Processing**, edited by W.T. Rhodes, J.R. Fienup and B.E.A. Saleh, SPIE vol. 373, pp.161-178 (1984).

R.J. Marks II and S.M. Tseng "Effect of sampling on closed form bandlimited signal interval interpolation", *Applied Optics*, vol. 24, pp.763-765 (1985); Erratum, vol. 24, p.2490 (1985).

A. Papoulis "A new method of image restoration", *Joint Services Technical Activity Report 39*, 1973-74.

A. Papoulis "A new algorithm in spectral analysis and bandlimited signal extrapolation", *IEEE Transactions on Circuits and Systems*, vol. CAS-22, pp.735-742 (1975).

A. Papoulis, **Signal Analysis**, McGraw-Hill, New York, 1977.

C. Pask "Simple optical theory of optical super-resolution", *Journal of the Optical Society of America*, vol. 66, pp. 68-69 (1976).

D. Slepian and H.O. Pollak "Prolate spheroidal wave functions Fourier analysis and uncertainty I", *Bell System Technical Journal*, vol. 40, pp.43-63 (1961).

H. Stark, editor, **Image Recovery: Theory and Application**, Academic Press, Orlando, 1987.

D.C. Youla "Generalized image restoration by method of alternating orthogonal projections", *IEEE Transactions on Circuits and Systems*, vol. CAS-25, pp.694-702 (1978).

D.C. Youla and H. Webb "Image restoration by the method of convex projection: Part 1 - theory", *IEEE Trans. Med. Imaging*, vol.MI-1, pp.81-94, 1982.

Appendix: Solution to Selected Problems

Solutions, Chapter 2

2.1 Since
$$X(0) = \int_{-\infty}^{\infty} x(t)dt,$$

the transforms in Table 2-2 can be used. The functions $\Pi(t), \operatorname{sinc}(t), \Lambda(t)$ and $\operatorname{jinc}(t)$ all have unit area. $J_0(2\pi t)$ has an area of $1/\pi$.

2.2 (a) $\delta(t) = \lim_{A\to\infty} A\Lambda(At)$.

   (b) $\delta(t) = \lim_{A\to\infty} A\operatorname{jinc}(At)$.

   (c) $\delta(t) = \lim_{A\to\infty} A\operatorname{sinc}^2(At)$.

   In each case, the limit of the transform approaches one. For example

$$A\operatorname{jinc}(At) \longleftrightarrow \sqrt{1-(u/A)^2}\Pi(\frac{u}{2A})$$
$$\longrightarrow 1 \text{ as } A \to \infty.$$

2.3 Selected solutions

   (a) Derivative

$$\left(\frac{d}{dt}\right)^n x(t) = \int_{-\infty}^{\infty} X(u)\left(\frac{d}{dt}\right)^n e^{j2\pi ut}du$$
$$= \int_{-\infty}^{\infty} (j2\pi u)^n X(u)e^{j2\pi ut}du.$$

(b) Convolution

$$x(t) * h(t) \longleftrightarrow \int_{-\infty}^{\infty} [\int_{-\infty}^{\infty} x(\tau)h(t-\tau)d\tau]e^{-j2\pi ut}dt.$$

Reverse integration orders and applying the shift theorem gives

$$x(t) * h(t) \longleftrightarrow H(u)\int_{-\infty}^{\infty} x(t)e^{-j2\pi ut}dt.$$

(c) Inversion

$$\int_{-\infty}^{\infty} X(v)e^{j2\pi vt}dv \longleftrightarrow \int_{-\infty}^{\infty}[\int_{-\infty}^{\infty} X(v)e^{j2\pi vt}dv]e^{-j2\pi ut}du.$$

Reverse integration order and note

$$\int_{-\infty}^{\infty} e^{j2\pi \xi t}dt = \delta(\xi).$$

Proof results from sifting property.

(d) Duality

$$X(t) \longleftrightarrow \int_{-\infty}^{\infty} X(t)e^{j2\pi(-u)t}dt$$

$$= x(-u).$$

2.4 (a)

$$\delta(t) \longleftrightarrow \int_{-\infty}^{\infty} \delta(t)e^{-j2\pi ut}dt$$

$$= 1$$

by the sifting property of the Dirac delta

(b)

$$\Pi(t) \longleftrightarrow \int_{-1/2}^{1/2} e^{-j2\pi ut}dt$$

$$= \frac{-1}{\pi t}\frac{1}{j2}e^{-j2\pi ut}\Big|_{-1/2}^{1/2}$$

$$= \text{sinc}(u).$$

(c) Use (2.14) and the convolution property in Table 2.

(d) Use (2.20). The result follows from linearity.

(e) Use (2.15) with $\nu = 1$ and apply (2.18).

2.5 (a) $g(t) \approx a(t - t_0)$ around the neighborhood $t = t_0$. Since $\delta(t)$ is nonzero only when $\xi = 0$, we conclude

$$
\begin{aligned}
\delta(g(t)) &= \delta(a(t - t_0)) \\
&= \frac{1}{|a|}\delta(t - t_0).
\end{aligned}
$$

(b)

$$
g(t) = \ln(t/b); \qquad t_0 = b
$$
$$
g'(t) = 1/t \longrightarrow a = 1/b
$$

and

$$
\delta(\ln(t/b)) = b\delta(t - b).
$$

(c) The argument has an infinite number of zero crossings:

$$
\sin(\pi t_n) = 0 \quad ; t_n = 0,\ \pm 1,\ \pm 2,\ \dots.
$$

Since

$$
\begin{aligned}
g'(t_n) &= \pi \cos(\pi t_n) \\
&= (-1)^n \pi \\
&= a_n,
\end{aligned}
$$

we have

$$
\begin{aligned}
\pi\delta(\sin(\pi t)) &= \pi \sum_{n=-\infty}^{\infty} \frac{1}{|a_n|}\delta(t - t_n) \\
&= \text{comb}(t).
\end{aligned}
$$

2.6 (a) From duality

$$
\text{sinc}(t) \longleftrightarrow \Pi(u)
$$

2.7 (a) (i) Use Parseval's theorem with $x(t) = \text{sinc}^2(t)$:

$$\int_{-\infty}^{\infty} \text{sinc}^4(t)dt = \int_{-1}^{1} \Lambda^2(u)du$$
$$= 2\int_{0}^{1}(1-u)^2 du$$
$$= 2/3.$$

(ii) Same approach, but $x(t) = \text{jinc}(t)$:

$$\int_{-\infty}^{\infty} \text{jinc}^2(t)dt = \int_{-1}^{1}(1-u^2)du$$
$$= 4/3.$$

(iii) Parseval's theorem gives

$$\int_{-\infty}^{\infty} J_0^2(2\pi t)dt = \frac{1}{\pi^2}\int_{-1}^{1}\frac{du}{1-u^2}$$
$$= \infty.$$

(b) Using the power theorem and (2.19) gives

$$\int_{-\infty}^{\infty} \text{sinc}(t)\text{jinc}(t)dt = \int_{-1/2}^{1/2}\sqrt{1-u^2}du$$
$$= \frac{\sqrt{3}}{4} + \frac{\pi}{6}.$$

2.8 (a) Use the Poisson Sum Formula:

$$\sum_{n=-\infty}^{\infty} \text{sinc}(t-n) = \sum_{n=-\infty}^{\infty} \Pi(n)e^{j2\pi nt}$$
$$= 1$$

since $\Pi(n) = \delta[n]$.

(b) Same result since $\Lambda(n) = \delta[n]$.

(c) Same result since $\sqrt{1-n^2}\Pi(n/2) = \delta[n]$.

(d) $\text{sinc}(at) \longleftrightarrow \Pi(u/a)/a$. Thus

$$\sum_{n=-\infty}^{\infty} \text{sinc}[a(t-n)] = \frac{1}{a}\sum_{n=-\infty}^{\infty} \Pi\left(\frac{n}{a}\right) e^{j2\pi nt}$$

$$= \frac{1}{a}\sum_{n=-N}^{N} e^{j2\pi nt}$$

where N is the greatest integer not exceeding $a/2$. Using a geometric series:

$$\sum_{n=-\infty}^{\infty} \text{sinc}[a(t-n)] = \frac{\sin[(2N+1)\pi t]}{a\sin(\pi t)}$$

Part (a) is a special case for $a = 1$ ($N = 0$).

$$\sum_{n=-\infty}^{\infty} \text{sinc}[a(t-n)] = \frac{1}{a}[1 + 2\sum_{n=1}^{N} \cos(2\pi nt)].$$

Part (a) is special case for $a = 1 (N = 0)$.

(e) Use (2.24) with $T = 2$

$$\sum_{n=-\infty}^{\infty} \text{jinc}(2n) = \frac{1}{2}\sum_{n=-\infty}^{\infty} \sqrt{1-(n/2)^2}\Pi(n/4)$$

$$= (1+\sqrt{3})/2.$$

(f) Use (2.24) with $T = 1/4$:

$$\sum_{n=-\infty}^{\infty} \text{jinc}^2(n/4) = 4\sum_{n=-\infty}^{\infty} C(4n)$$

where

$$C(u) = \sqrt{1-u^2}\Pi(u/2) * \sqrt{1-u^2}\Pi(u/2)$$

$$= C(u)\Pi(u/4).$$

Thus :

$$\sum_{n=-\infty}^{\infty} \text{jinc}^2(n/4) = 4C(0)$$

where

$$C(0) = \int_{-1}^{1}(1-u^2)du$$
$$= 4/3.$$

(g) The $n > 0$ and $n < 0$ terms cancel. The $n = 0$ term is zero.

2.9 (a) $(\frac{d}{dt})\mathrm{sinc}(t) = 0 \implies \theta = \tan(\theta); \theta = \pi t$. The nth extrema location, θ_n, can be founded iteratively by $\theta_n[m] \longrightarrow \theta_n$ as $m \longrightarrow \infty$ with

$$\theta_n[m+1] = n\pi + \arctan(\theta_n[m]) \ ; \ n = 0, \pm 1, \pm 2, \ldots.$$

The first few locations and the corresponding extrema are listed in Table A.1.

(b) $\theta_n \longrightarrow n + 1/2$ as $n \longrightarrow \infty$ and

$$\mathrm{sinc}(t_n) \longrightarrow \frac{(-1)^n}{\pi(n+\frac{1}{2})}.$$

(To justify, compare plots of θ and $\tan(\theta)$ vs. θ).

2.10 The Fourier series is

$$y(t) = \alpha \sum_{n=-\infty}^{\infty} \mathrm{sinc}(\alpha n)e^{j2\pi n t/T}.$$

Thus

$$y(\tau/2) = \alpha \sum_{n=-\infty}^{\infty} \mathrm{sinc}(\alpha n) \, \cos(n\pi\alpha).$$

Write sinc as $\sin(\pi x)/(\pi x)$ and use a trigonometric identity:

$$y(\tau/2) = \alpha \sum_{n=-\infty}^{\infty} \mathrm{sinc}(2\alpha n).$$

We evaluate using (2.24) which, for $x(t) = \mathrm{sinc}(t)$, can be written

$$\alpha \sum_{n=-\infty}^{\infty} \mathrm{sinc}(2\alpha n) = \frac{1}{2} \sum_{n=-\infty}^{\infty} \Pi(\frac{n}{2\,\alpha})$$
$$= 1/2.$$

n	θ_n	$\text{sinc}(\theta_n)$
1	4.4934	-0.21723
2	7.7253	0.12837
3	10.9041	-0.09133
4	14.0662	0.07091
5	17.2208	-0.05797
6	20.3713	0.04903
7	23.5194	-0.04248
8	26.6661	0.03747
9	29.8116	-0.03353
10	32.9564	0.03033
11	36.1006	-0.02769
12	39.2444	0.02547

Table A.1: The first few extrema of $\text{sinc}(t)$.

2.11   Since the integrand in (2.24) is even

$$\frac{J_\nu(2\pi t)}{(2t)^\nu} = \frac{(\pi/2)^\nu}{\sqrt{\pi}\Gamma(\nu + \frac{1}{2})} \int_{-1}^{1} (1 - u^2)^{\nu - \frac{1}{2}} \cos(2\pi ut) \, du.$$

Thus we have the transform pair

$$\frac{J_\nu(2\pi t)}{(2t)^\nu} \longleftrightarrow \frac{(\pi/2)^\nu}{\sqrt{\pi}\Gamma(\nu + \frac{1}{2})}(1 - u^2)^{\nu - \frac{1}{2}}\Pi(u/2).$$

The entries in Table 2-2 for $\nu = 0$, 1 follow immediately using (2.16).

2.12   With attention to Fig.A.1:

$$z_T(t) = \Pi(\frac{t - \frac{T}{4}}{T/2}) - \Pi(\frac{t + \frac{T}{4}}{T/2}).$$

Thus

$$Z_T(u) = -j \, \text{sinc}(\frac{Tu}{2}) \, \sin(\frac{\pi ut}{2}).$$

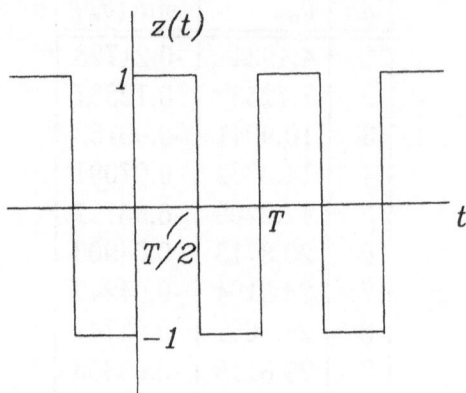

Figure A.1

The Fourier coefficients follow as:

$$c_n = \frac{1}{T}Z_T(n/T)$$
$$= -j\,\text{sinc}(n/2)\sin(\pi n/2).$$

Thus, since $z(t)$ is real :

$$z(t) = \sum_{n=-\infty}^{\infty} -j\text{sinc}(n/2)\,\sin(\pi n/2)\,e^{j2\pi nt/T}$$
$$= \sum_{n=-\infty}^{\infty} \text{sinc}(n/2)\,\sin(\pi n/2)\,\sin(\frac{2\pi nt}{T}).$$

The truncated series is

$$z_N(t) = \sum_{n=-N}^{N} \text{sinc}(n/2)\,\sin(\pi n/2)\,\sin(\frac{2\pi nt}{T}).$$

Extrema come from:

$$0 = \frac{d}{dt}z_N(t)$$
$$= \frac{2\pi}{T}\sum_{n=-N}^{N} \text{sinc}(n/2)\,\sin(\pi n/2)\,n\,\cos(\frac{2\pi nt}{T})$$

$$= \frac{4}{T} \sum_{n=-N}^{N} \sin^2(\pi n/2) \, \cos(\frac{2\pi nt}{T}).$$

But $\sin^2(\pi n/2) = [1 - (-1)^n]/2$ and

$$0 = \frac{2}{T} \sum_{n=-N}^{N} [1 - (-1)^n] \, \cos(\frac{2\pi nt}{T})$$

$$= \frac{2}{T} \Re \sum_{n=-N}^{N} [1 - (-1)^n] \, e^{j2\pi nt/T}.$$

Consider the geometric series:

$$S = \sum_{n=-N}^{N} a^n = a^{-N} + a^{-N+1} + \ldots + a^N.$$

Thus

$$aS = a^{-N+1} + \ldots + a^N + a^{N+1}.$$

Substituting:

$$S = \frac{a^{N+\frac{1}{2}} - a^{-(N+\frac{1}{2})}}{a^{1/2} - a^{-1/2}}.$$

For $a = -e^{-j2\pi t/T} = -e^{j\theta}$

$$\text{Sum} = \frac{(-1)^N \, \cos(N + \frac{1}{2})\theta}{\cos(\theta/2)}.$$

For $a = e^{j\theta}$

$$\text{Sum} = \frac{\sin(N + \frac{1}{2})\theta}{\sin(\theta/2)}.$$

Thus

$$0 = \frac{\sin(N + \frac{1}{2})\theta}{\sin(\theta/2)} - \frac{(-1)^N \, \cos(N + \frac{1}{2})\theta}{\cos(\theta/2)}$$

or, for N even:

$$\tan(N + \frac{1}{2})\theta = \tan(\theta/2).$$

Therefore

$$(N + \frac{1}{2})\theta = \frac{\theta}{2} + p\pi \; ; \quad p = 0, \pm 1, \pm 2, \ldots$$

or

$$\theta = \frac{2\pi t}{T} = \frac{p\pi}{N} \implies t = \frac{pT}{2N}.$$

The first extrema is at $p = 1$ corresponding to $t = T/2N$. Substituting gives

$$z_N(\frac{T}{2N}) = \sum_{n=-N}^{N} \frac{\sin^2(\pi n/2)}{\pi n/2} \sin(\pi n/N)$$

$$= 2 \sum_{n=1}^{N} \frac{\sin^2(\pi n/2)}{\pi n/2} \sin(\pi n/N)$$

$$= \frac{2}{\pi} \sum_{n=1}^{N} \frac{1 - (-1)^n}{n} \sin(\pi n/N).$$

Let $2m + 1 = n$

$$z_N(\frac{T}{2N}) = \frac{4}{\pi} \sum_{m=0}^{N/2} \frac{\sin \frac{\pi(2m+1)}{N}}{2m + 1}.$$

Define

$$h(t) = \sum_{m=0}^{M} \frac{\sin(2m + 1)t}{2m + 1} \; ; \quad M = N/2$$

so that

$$z_N(\frac{T}{2N}) = \frac{4}{\pi} h(\pi/N).$$

Note that $h(0) = 0$. Now

$$\frac{dh(t)}{dt} = \sum_{m=0}^{N} \cos(2m + 1)t$$

$$= \frac{\sin 2(M + 1)t}{2 \sin(t)}.$$

Thus:

$$h(t) = \int_0^t \frac{\sin 2(M + 1)\tau}{2 \sin(\tau)} d\tau$$

and
$$z_N\left(\frac{T}{2N}\right) = \frac{4}{\pi} \int_0^{\pi/N} \frac{\sin(N+1)\tau}{2\sin(\tau)}\, d\tau.$$

Since the interval of integration gets smaller and smaller, $\sin\tau \longrightarrow \tau$ and as $N \to \infty$

$$z_N\left(\frac{T}{2N}\right) \longrightarrow \frac{4}{\pi}\int_0^{\pi/N} \frac{\sin(N+2)\tau}{2\tau}\, d\tau$$

$$= \frac{2}{\pi}\int_0^{\pi+\frac{2\pi}{N}} \frac{\sin(\xi)}{\xi}\, d\xi.$$

Since $\mathrm{Si}(\pi) = 1.8519370$, we conclude that, as $N \to \infty$,

$$\Delta \longrightarrow 1 - \tfrac{2}{\pi}\mathrm{Si}(\pi)$$
$$= 0.1789797.$$

2.13 (a) $\overline{\xi^2} = R_\xi(0) = \int_{-\infty}^{\infty} S_\xi(u)du.$

(b) $\overline{\xi^2} = R_\xi[0].$

Note that (2.28) is a Fourier series with period 1. Thus

$$\overline{\xi^2} = \int_T S_\xi(u)du$$

where integration is over any period.

2.14 These derivations can be found in most any introductory text on stochastic processes.

2.15 Same as 2.14.

2.16 (a) From (2.26)

$$R_\xi^*(\tau - t) = E[\xi(\tau)\,\xi^*(t)]^*$$
$$= R_\xi(t - \xi).$$

(b)

$$S_\xi^*(u) = \int_{-\infty}^{\infty} R_\xi^*(t)\, e^{j2\pi ut}\, dt$$
$$= \int_{-\infty}^{\infty} R_\xi^*(-\tau)\, e^{-j2\pi u\tau}\, d\tau$$
$$= \int_{-\infty}^{\infty} R_\xi(\tau)\, e^{-j2\pi u\tau}\, d\tau$$
$$= S_\xi(u).$$

	$(0,1)$	$(1,\infty)$
A	4	∞
E	∞	2

Table A.2

2.17 Using the shift theorem

$$\mathrm{comb}(t - 1/2) \longleftrightarrow \mathrm{comb}(u)\, e^{-j\pi u}$$
$$= \textstyle\sum_{n=-\infty}^{\infty} \delta(u - n)\, e^{-j\pi n}$$
$$= \textstyle\sum_{n=-\infty}^{\infty} (-1)^n\, \delta(u - n).$$

Separating the sum into even and odd n completes the problem.

2.18 (a) See Table A.2.

(b) on the interval $(0, 1)$, $x(t)$ has finite area and energy. On the interval $(0, \infty)$, the area is infinite and the energy is finite.

(c) $A < \infty \implies F < \infty$. This follows from the proof that the space of l_1 sequences is subsumed in l_2 [Luenburger: Naylor and Sell]. The converse is not true. Consider

$$x(t) = \mathrm{sinc}(Bt).$$

Clearly, $E = 1/B$. However, since

$$\left| x\left(\frac{n}{2B}\right) \right| = \begin{cases} 1 & ; \ n = 0 \\ \frac{2}{\pi|n|} & ; \ \text{odd } n \\ 0 & ; \ \text{otherwise} \end{cases}$$

we have the divergent series

$$A = 1 + \frac{2}{\pi} \sum_{k=0}^{\infty} \frac{1}{2k + 1} = \infty.$$

2.19 (a)

$$|x(t)| = \left| \int_{-\infty}^{\infty} X(u)\, e^{j2\pi ut}\, du \right|$$
$$\leq \int_{-\infty}^{\infty} |X(u)|\, du$$
$$= A.$$

Thus, $C = A$. Note, then, that $x(t)$ is bounded if $X(u)$ has finite area.

(b) A counterexample is $x(t) = \text{sgn}(t)$.

2.20

$$|x(t)|^2 = |\int_{-B}^{B} X(u)\, e^{j2\pi ut}\, du|^2$$
$$\leq 2B \int_{-B}^{B} |X(u)|^2\, du$$
$$= 2BE.$$

2.21 (a) Yes, because applying Parseval's theorem to the derivative theorem gives

$$E_p = \int_{-\infty}^{\infty} |x^{(p)}(t)|^2\, dt$$
$$= \int_{-B}^{B} |(j2\pi u)^p\, X(u)|^2\, du.$$

Since $(u/B)^{2p} < 1$ over the interval $|u| < B$,

$$E_p = (2\pi B)^{2p} \int_{-B}^{B} (u/B)^{2p}\, |X(u)|^2\, du$$
$$\leq (2\pi B)^{2p} \int_{-B}^{B} |X(u)|^2\, du$$
$$= (2\pi B)^{2p}\, E$$

where E is the energy of $x(t)$. Also, $x^{(p)}(t)$ is clearly bandlimited.

(c) Yes:

$$|x^{(p)}(t)| = |\int_{-B}^{B} (j2\pi u)^p\, X(u)\, e^{j2\pi ut}\, du|$$
$$\leq \int_{-B}^{B} |2\pi u|^p\, |X(u)|\, du$$
$$= (2\pi B)^p \int_{-B}^{B} |u/B|^p\, |X(u)|\, du$$
$$\leq (2\pi B)^p\, A$$

where A is the area of $X(u)$.

2.22   The series converges absolutely if

$$S(t) \equiv \sum_{m=0}^{\infty} \frac{|t - \tau|^m}{m!} |x^{(m)}(\tau)| < \infty.$$

From the derivative theorem for Fourier transforms,

$$
\begin{aligned}
S(t) &= \sum_{m=0}^{\infty} \frac{|t-\tau|^m}{m!} \left| \int_{-B}^{B} (j2\pi u)^m X(u) e^{j2\pi ut} \, du \right| \\
&\leq \sum_{m=0}^{\infty} \frac{|t-\tau|^m}{m!} \int_{-B}^{B} (2\pi u)^{2m} \, du]^{1/2} \\
&\quad \cdot [\int_{-B}^{B} |X(u)|^2 \, du]^{1/2}
\end{aligned}
$$

where, in the second step, we have used Schwarz's inequality. Since $x(t)$ has finite energy,

$$E = \int_{-B}^{B} |X(u)|^2 \, du$$

is finite. Thus

$$
\begin{aligned}
S(t) &\leq \quad \sqrt{2B} \; E \sum_{m=0}^{\infty} \frac{(2\pi B |t-\tau|^2)^m}{m! \sqrt{2m+1}} \\
&< \sqrt{2B} \; E \sum_{m=0}^{\infty} \frac{(2\pi B |t-\tau|)^m}{m!} \\
&= \sqrt{2B} \; E \; e^{2\pi B |t-\tau|}.
\end{aligned}
$$

This bound is finite for all finite t and τ.

2.23   (a)   From the derivative theorem

$$
\begin{aligned}
|x^{(M)}(t)|^2 &= \left| \int_{-B}^{B} (j2\pi u)^M X(u) \, du \right|^2 \\
&\leq 2(2\pi)^{2M} E \int_{-B}^{B} u^{2M} \, du
\end{aligned}
$$

which, when evaluated, gives (2.39).

(b)   Clearly

$$
\begin{aligned}
|x(t+\tau) - x(t)| &= \left| \int_{t}^{t+\tau} x'(\xi) \, d\xi \right| \\
&\leq \int_{t}^{t+\tau} |x'(\xi)| \, d\xi \\
&\leq \sqrt{\frac{(2\pi B)^3 E}{3}} \int_{t}^{t+\tau} d\xi
\end{aligned}
$$

which, when evaluated, gives (2.40).

2.24 Define

$$E_f = \int_{-\infty}^{\infty} |f(t)|^2 \, dt.$$

Then, applying Schwarz's inequality to the integral gives

$$|g(t)|^2 \leq E_f \, E_h(t).$$

Note that Exercise 2.19 is a special case for the Fourier inversion integral over the interval $|u| \leq B$.

Solutions, Chapter 3

3.1   The Fourier dual of the Poisson sum formula is

$$2B \sum_{n=-\infty}^{\infty} X(u - 2nB) = \sum_{n=-\infty}^{\infty} x(\frac{n}{2B}) e^{-j\pi nu/B}.$$

Since $X(u) = X(u) \, \Pi(u/2B)$, we multiply both sides by $\Pi(\frac{u}{2B})$ and inverse transform. The sampling theorem series results.

3.2   Use the Fourier dual of the Poisson sum formula again:

$$\sum_{n=-\infty}^{\infty} X(u - \frac{n}{T}) = T \sum_{n=-\infty}^{\infty} x(nT) \, e^{-j2\pi nuT}.$$

For $B < 1/T < 2B$, the sum on the left will be the overlapping aliased version in Fig.A.2. Note that no spectra overlap the zeroth order spectrum at $u = 0$. Thus

$$\sum_{n=-\infty}^{\infty} X(u - \frac{n}{T}) \,|_{u=0} = X(0) \; ; B < 1/T < 2B$$

and

$$\int_{-\infty}^{\infty} x(t) \, dt = T \sum_{n=-\infty}^{\infty} x(nT) \; ; T < 1/B.$$

3.3   If $x(t)$ is bandlimited, so is $x(t + \alpha)$. Thus

$$x(t + \alpha) = \sum_{n=-\infty}^{\infty} x(\frac{n}{2B} + \alpha) \, \text{sinc}(2Bt - n).$$

Substitute $t - \alpha$ for t and we're done.

3.4   Since

$$(-j2\pi t)^m \longleftrightarrow \delta^{(m)}(t)$$

we use (3.4) and ask the question :

$$t^m \stackrel{?}{=} \frac{1}{\pi(2B)^m} \sin(2\pi Bt) \sum_{n=-\infty}^{\infty} \frac{(-1)^n \, n^m}{2Bt - n}.$$

The result is clearly a divergent series for $m > 1$.

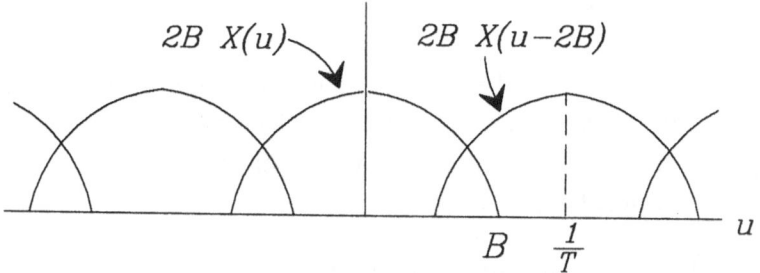

Figure A.2

3.5

$$\hat{s}(t) = \sum_{n=-\infty}^{\infty} y(nT)\, \delta(t - nT).$$

Thus

$$\hat{S}(u) = \sum_{n=-\infty}^{\infty} y(nT)\, e^{-j2\pi nuT}$$

$\hat{S}(u)$ is periodic with Fourier coefficients:

$$y(nT) = T \int_{-1/T}^{1/T} \hat{S}(u)\, e^{j2\pi nuT}\, du.$$

Thus

$$T \int_{-1/T}^{1/T} \hat{S}(u)\, du = y(0).$$

From the inversion formula:

$$y(t) = \int_{-\infty}^{\infty} Y(u)\, e^{j2\pi ut}\, du.$$

Thus

$$y(0) = \int_{-\infty}^{\infty} Y(u)\, du$$

and our exercise is complete.

3.6

$$\int_{-\infty}^{\infty} x(t)\, y^*(t)\, dt = \frac{1}{2B} \sum_{n=-\infty}^{\infty} x(\frac{n}{2B})\, y^*(\frac{n}{2B}).$$

3.7 The new signal's spectrum is no longer Hermetian. The signal, therefore, is complex. Although the sampling rate is reduced by a factor of a half, each sample now requires two numbers.

3.8 The resulting low pass kernel is

$$k(u;t) = \exp(-j2\pi ut)\, \Pi(\frac{u}{2B}).$$

We are thus assured that $g(u)$, the Fourier transform of $f(t)$, is zero outside of the interval $|u| \leq B$. Since this is the definition of a bandlimited function, the result is of little use. Note that Fourier inversion of (3.23), as we would expect, results in the cardinal series.

Solutions, Chapter 4

4.1 (a)

$$y(t; C) = D C^2 \sum_{n=-\infty}^{\infty} x(nT) \operatorname{sinc}^2[C(t - nT)]$$

$$y(t; W) = D W^2 \sum_{n=-\infty}^{\infty} x(nT) \operatorname{sinc}^2[W(t - nT)]$$

$$x(t) = \sum_{n=-\infty}^{\infty} x(nT) k_r(t - n/2W)$$

where

$$k_r(t) = D [C^2 \operatorname{sinc}^2(Ct) - B^2 \operatorname{sinc}^2(Bt)]$$
$$\longleftrightarrow \quad K_r(u) = D [C\Lambda(u/C) - B\Lambda(u/B)].$$

For $|u| \le B$,

$$K_r(u) = D [C\{1 - \frac{|u|}{C}\} - B \{1 - \frac{|u|}{B}\}]$$
$$= T ; \quad D = \frac{1}{2W(C - B)}$$

and $x(t)$ is recovered.

(b) Here,

$$k_r(t) = D C^2 \operatorname{sinc}(Ct)$$
$$\longleftrightarrow \quad K_r(u) = K(\frac{u}{2B}) = D C\Lambda(u/C)$$

and

$$H(\frac{u}{2B}) = \Pi(\frac{u}{2B}) / K(\frac{u}{2B})$$
$$= \frac{1}{D(C - |u|)} \Pi(\frac{u}{2B}).$$

Notice, for $|u| \le B$,

$$1/DC \le H(\frac{u}{2B}) \le \frac{1}{D(C - B)}.$$

4.2 (a) $K(\frac{u}{2B}) = 1/(a + j2\pi u)$ and

$$H(\frac{u}{2B}) = (a + j2\pi u)\ \Pi(\frac{u}{2B}).$$

This inverse filter is shown in Fig.A.3.[2]

(b)

$$K(\frac{u}{2B}) = \frac{2a}{a^2 - (j2\pi u)^2}$$

and

$$H(\frac{u}{2B}) = [\frac{a}{2} - \frac{1}{2a}\ (j2\pi u)^2\ ]\ \Pi(\frac{u}{2B}).$$

The inverse filter is shown in Fig.A.4.

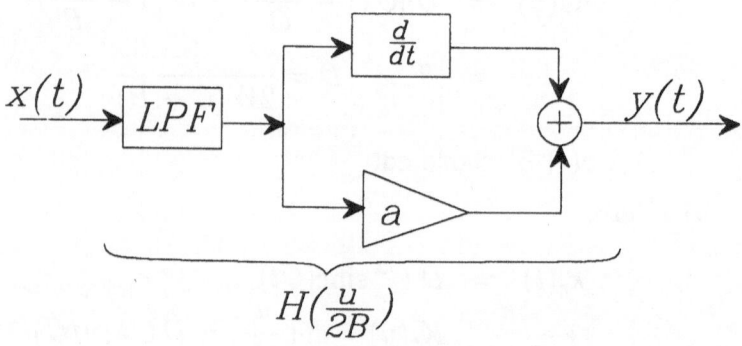

Figure A.3

4.3 (b) $H(u) = -j\ \text{sgn}(u)$. Since $1/\text{sgn}(u) = \text{sgn}(u)$,

$$K(u) = \frac{j}{2B}\ \text{sgn}(u)\ \Pi(\frac{u}{2B})$$

---

[2]Note: the LPF should be used first to avoid differentiating discontinuities.

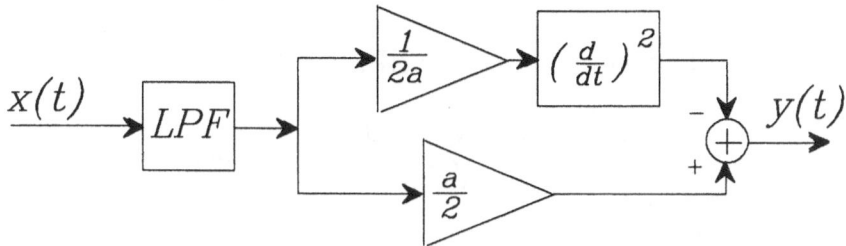

Figure A.4

and

$$k(t) = \frac{j}{2B} \int_{-B}^{B} \text{sgn}(u) \left[j \, \sin(2\pi ut) \right] du$$

$$= \frac{-1}{B} \int_{0}^{B} \sin(2\pi ut) \, du$$

$$= -\pi Bt \, \text{sinc}^2(Bt).$$

Thus, if $g(t)$ is the Hilbert transform of $f(t)$ with bandwidth B, then

$$f(t) = \frac{1}{\pi} \sum_{n=-\infty}^{\infty} g(\frac{n}{2B}) \frac{(-1)^n \, \cos(2\pi Bt) - 1}{2Bt - n}.$$

4.5 Doing so gives $\vec{\hat{g}} = \mathbf{S} \, \vec{x}$ or

$$\sum_{n \notin M} \{\delta[n - m] - r\text{sinc}(r(n - m))\} x(\frac{n}{2W})$$

$$= r \sum_{n \in M} x(\frac{n}{2W}) \, \text{sinc}(r(n - m))$$

where $\vec{\hat{g}}$ has elements $\{\hat{g}(\frac{n}{2B}) \mid n \in M\}$ and

$$\hat{g}(nT) = \sum_{n \notin M} \{\delta[n - m] - r\text{sinc}(r(n - m))\} x(\frac{n}{2W})$$

can be found from the known data.

4.6

$$f(nT) = \begin{cases} \frac{4(-1)^{n/2}}{\pi(1-n^2)} & ; \text{ even } n \\ 0 & ; \text{ odd } n, \ n \neq \pm 1 \end{cases}$$

Therefore, for $T = 1/2W$,

$$g(\pm T) = \frac{1}{2} \sum_{\text{even } n} \frac{4(-1)^{n/2}}{\pi(1-n^2)} \operatorname{sinc}(\frac{\pm 1 - n}{2}).$$

Note $g(T) = g(-T)$. Using the $+$ gives

$$\operatorname{sinc}(\frac{n-1}{2}) = \frac{2(-1)^{n/2}}{\pi(1-n)} \quad ; \text{even } n$$

and

$$\begin{aligned} g(\pm T) &= (2/\pi)^2 \sum_{\text{even } n} \frac{1}{(1-n^2)(1-n)} \\ &= (2/\pi)^2 [1 + 2\sum_{m=1}^{\infty} \frac{1}{[1-(2m)^2]^2} \end{aligned}$$

where we have let $2m = n$. Numerically,

$$g(\pm T) = 1/2.$$

Now,

$$\mathbf{H} = \begin{bmatrix} -1/2 & 0 \\ 0 & 1/2 \end{bmatrix}.$$

Thus

$$f(\pm T) = 1.$$

Note: The function these samples were taken from is

$$f(t) = \operatorname{sinc}(2Bt - \frac{1}{2}) + \operatorname{sinc}(2Bt + \frac{1}{2}).$$

4.8 The samples are taken from the signal

$$f(t) = \frac{d}{dt} \operatorname{sinc}(t) = \frac{\cos(\pi t) - \operatorname{sinc}(t)}{t}.$$

Thus $f(0) = 0$. (Note $r = 1/2$). Using (4.7) :

$$x(0) \;=\; -2 \sum_{\substack{\text{even } n \\ n \neq 0}} \frac{(-1)^{n/2}}{n} \;-\; \frac{4}{\pi} \sum_{\text{odd } n} \frac{(-1)^{\frac{n+1}{2}}}{n^2}$$

$$=\; 0$$

since both summands are odd.

4.11 (a) For $M = 1$ and $\alpha = 0$, (4.50) becomes

$$\Delta(u) = j2\pi B.$$

Equation (4.51) becomes

$$K_1(u) \;=\; \frac{1}{j2\pi B^2} \left[\Pi(\frac{u}{B} - \frac{1}{2}) - \Pi(\frac{u}{B} + \frac{1}{2}) \right]$$

$$=\; \frac{1}{j2\pi B^2} \, \Pi(\frac{u}{B}) * \left[\delta(u - \frac{B}{2}) - \delta(u + \frac{B}{2}) \right].$$

Thus

$$k_1(t) \;=\; \frac{1}{\pi B} \, \text{sinc}(Bt) \, \sin(\pi Bt)$$

$$=\; \frac{\sin^2(\pi Bt)}{\pi^2 B^2 t}.$$

Equation (4.52) becomes

$$K_2(u) \;=\; \frac{1}{B^2} \left[-(u - B)\Pi(\frac{u}{B} - \frac{1}{2}) + (u + B)\,\Pi(\frac{u}{B} + \frac{1}{2}) \right]$$

$$=\; \frac{1}{B} \, \Lambda(u/B).$$

Thus

$$k_2(t) \;=\; \text{sinc}^2(Bt)$$

$$=\; \frac{\sin^2(\pi Bt)}{(\pi Bt)^2}.$$

(b) The interpolation is

$$f(t) = \sum_{n=-\infty}^{\infty} \left[f(\frac{n}{B}) \, k_1(t - \frac{n}{B}) + f'(\frac{n}{B}) \, k_2(t - \frac{n}{B}) \right].$$

Since $k_1(p/B) = \delta[n]$ and $k_2(p/B) = 0$, the equation reduces to an identity at $t = m/B$.
Differentiate

$$f'(t) = \sum_{n=-\infty}^{\infty} [f(\frac{n}{B}) \, k_1'(t - \frac{n}{B}) + f'(\frac{n}{B}) \, k_2'(t - \frac{n}{B})].$$

Clearly:

$$
\begin{aligned}
k_1'(t) &= \frac{d}{dt} \, \text{sinc}^2(Bt) \\
&= 2B \, \text{sinc}(Bt) \, d_1(Bt)
\end{aligned}
$$

and

$$k_1'(\frac{n}{B}) = 0$$

since $d_1(0) = 0$ and $\text{sinc}(n)$ is zero everywhere else. Note that

$$k_2(t) = t \, k_1(t),$$

Thus

$$k_2'(t) = t \, k_1'(t) + k_1(t)$$

and

$$
\begin{aligned}
k_2'(\frac{n}{B}) &= \text{sinc}^2(n) \\
&= \delta[n].
\end{aligned}
$$

The interpolation therefore also interpolates the derivative samples.

4.12 We have implicitly assumed that $C_{\pm N} = 0$. If $P = 2N + 1$ then, in the spectral replication, the Dirac delta at $u = N/T$ would be aliased by the shifted Dirac delta originally at $u = -N/T$.

4.13 We evaluate the Fourier coefficients in (4.70) with the familiar formula

$$c_n = \frac{1}{T} \int_{-T/2}^{T/2} x(\tau)\, e^{j2\pi n\tau/T}\, d\tau.$$

Substituting (4.71) followed by manipulation completes the problem.

4.14 Since $v(t)$ is real, $V(u)$ is Hermitian. Thus, as illustrated in Fig. A.5, $X(u)$ has a four fold symmetry. We can, therefore, hetrodyne the center frequency f_0 (rather than the lower frequency f_L to the origin). Let

$$
\begin{aligned}
y(t) &= x(t)\cos(2\pi f_0 t) \\
&= v(t)\cos^2(2\pi f_0 t) \\
&= \frac{1}{2}v(t) + \frac{1}{2}v(t)\,\cos(\pi f_0 t).
\end{aligned}
$$

A lowpass filter gives

$$z(t) = \frac{1}{2}v(t)$$

which can be sampled at the Nyquist rate of B.

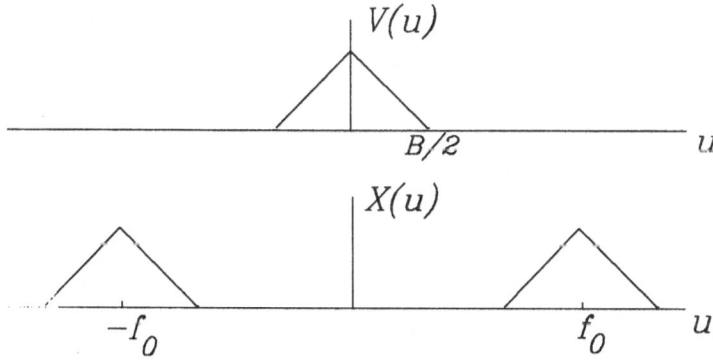

Figure A.5

4.15 Since $y(t) > 0$ and approaches zero for large $|t|$, no matter how small Δ, there exists some value of $|t|$ above which there will be no more samples. Thus, the number of samples is finite and the signal is not uniquely determined.

4.16 Clearly

$$k(n) = \text{sinc}(n/2) \; \cos[\pi(2N+1)n/2].$$

Except for $n = 0$, $\text{sinc}(n/2)$ is zero for even n. The cosine term is always zero for odd n. Therefore, (4.23) is satisfied.

4.17 Real N^{th} order polynomials.

4.18 We wish to compute the interpolation function corresponding to α_p. The Lagrangian kernel in (4.82) can be partitioned as

$$k_p(t) = a_p(t) \, b_p(t)$$

where

$$a_p(t) = \prod_{m \neq 0} \frac{t - (mT_N + \alpha_p)}{\alpha_p - (mT_N + \alpha_p)}$$

is due to sample locations located distances mT from α_p and the contribution due to the remaining terms is

$$b_p(t) = \prod_{\substack{q = 1 \\ q \neq p}} \prod_{m=-\infty}^{\infty} \frac{t - (mT_N + \alpha_p)}{\alpha_p - (mT_N + \alpha_p)}.$$

Using the product formula for $\sin(z)$ (Section 4.8),

$$a_p(t) = \text{sinc}[2B_N(t - \alpha_p)].$$

After factoring out the zero term, the m product in $b_p(t)$ can be written as

$$\frac{t + \alpha_q}{\alpha_p - \alpha_q} \prod_{m \neq 0} \frac{1 - \frac{t - \alpha_q}{mT_N}}{1 - \frac{\alpha_p - \alpha_q}{mT_N}}.$$

Expressing both products as sincs and simplifying reveals the resulting interpolation function to be identical to that in (4.47).

4.19 No. Lagrangian interpolation would result in the conventional cardinal series. Recall that Lagrangian interpolation results in an interpolation where only the sample value contributes to the interpolation at that point. Equation(4.2), on the other hand, usually has every sample value contributing to the sample at $t = m/2W$.

4.20 We write (4.2) as

$$x(t) = rx(0)\text{sinc}(2Bt) + r \sum_{n \neq 0} x(n/2B) \text{ sinc}(2Bt - rn).$$

Substitute (4.9) and simplify. The same result can be obtained by filtering (4.10).

4.22 The matrix to solve is (for $\frac{B}{3} < u < B$)

$$\begin{bmatrix} S_1(u) \\ S_2(u) \\ S_3(u) \end{bmatrix} = \frac{2B}{3} \begin{bmatrix} 1 & 1 & 1 \\ j2\pi u & j2\pi(u - \frac{2B}{3}) & j2\pi(u - \frac{4B}{3}) \\ (j2\pi u)^2 & [j2\pi(u - \frac{2B}{3})]^2 & [j2\pi(u - \frac{4B}{3})]^2 \end{bmatrix}$$

$$\times \begin{bmatrix} F(u) \\ F(u - \frac{2B}{3}) \\ F(u - \frac{4B}{3}) \end{bmatrix}.$$

Using the third order Vandermonde determinant

$$\Delta(u) = j2(\frac{4\pi B}{3})^3$$

and Cramer's rule, we have

$$\begin{aligned} F(u) &= \frac{1}{2} (\frac{3}{2B})^3 [(u - \frac{2B}{3})(u - \frac{4B}{3}) S_1(u) \\ &\quad + j(\frac{1}{2\pi})(\frac{3}{2B})^2 (u - B) S_2(u) \\ &\quad - (\frac{1}{8\pi^2})(\frac{3}{2B})^3 S_3(u) \; ; \quad \frac{B}{3} < u < B \end{aligned}$$

$$F(u - \frac{2B}{3}) = -(\frac{3}{2B})^3 u(u - \frac{4B}{3}) S_1(u)$$
$$-j(\frac{1}{\pi})(\frac{3}{2B})^3 (u - \frac{2B}{3}) S_2(u)$$
$$+(\frac{1}{4\pi^2})(\frac{3}{2B})^3 S_3(u) \ ; \quad \frac{B}{3} < u < B$$
$$F(u - \frac{4B}{3}) = \frac{1}{2}(\frac{3}{2B})^3 u(u - \frac{2B}{3}) S_1(u)$$
$$+j(\frac{1}{2\pi})(\frac{3}{2B})^3 (u - \frac{B}{3}) S_2(u)$$
$$-(\frac{1}{8\pi^2})(\frac{3}{2B})^3 S_3(u) \ ; \quad \frac{B}{3} < u < B.$$

To construct $F(u)$, we evaluate the equation

$$F(u) = K_1(u) \, S_1(u) + K_2(u) \, S_2(u) + K_3(u) \, S_3(u) \ ; -B < u < B$$

Solving gives

$$K_1(u) = \frac{1}{2}(\frac{3}{2B})^3 \, [(u - \frac{2B}{3})(u - \frac{4B}{3})\Pi(\frac{3u}{2B} - 1)$$
$$-2\,(u - \frac{2B}{3})(u + \frac{2B}{3})\Pi(\frac{3u}{2B})$$
$$+(u + \frac{2B}{3})(u + \frac{4B}{3})\Pi(\frac{3u}{2B} + 1)],$$
$$K_2(u) = j(\frac{1}{2\pi})(\frac{3}{2B})^2[(u - B)\Pi(\frac{3u}{2B} - 1)$$
$$-2u\Pi(\frac{3u}{2B}) + (u + B)\Pi(\frac{3u}{2B} + 1)]$$

and

$$K_3(u) = -(\frac{1}{8\pi^2})(\frac{3}{2B})^3[\Pi(\frac{3u}{2B} - 1)$$
$$-2\Pi(\frac{3u}{2B}) + \Pi(\frac{3u}{2B} + 1)].$$

With a bit of work we inverse Fourier transform to the interpolation functions in (4.57).

Solutions, Chapter 5

5.1 The explanation is unreasonable because the noise level on a bandlimited signal could, in the limit, be reduced to zero. The resolution is that any physical continuous noise has a finite correlation length. Thus, as the samples are taken closer and closer, the noise eventually must become correlated and the white noise assumption is violated. For continuous white noise, all noise samples are uncorrelated. But the noise level is infinite.

5.3 $\overline{\Psi^2} = c_0$.

5.4 Consider a power spectral density that is nonzero only over the interval $B < |u| < W$. Interpolating and filtering would yield a zero NINV.

5.6 Clearly

$$E_z = \sum_{n=-\infty}^{\infty} [x^2(\widetilde{n/2W}) - x(\widetilde{n/2W})^2] \int_{-\infty}^{\infty} k^2(t)\, dt.$$

From (5.73) and Parseval's theorem,

$$\int_{-\infty}^{\infty} k^2(t)\, dt = \frac{1}{(2W)^2} \int_{-B}^{B} |\Phi_\theta(u)|^{-2} du.$$

The case of $y(t)$ can be obtained by $k(t) = r\,\text{sinc}(2Bt)$. Since

$$r^2 \int_{-\infty}^{\infty} \text{sinc}^2(2Bt)\, dt = \frac{1}{(2W)^2} \int_{-B}^{B} du$$

$$= \frac{r}{2W} \le \int_{-\infty}^{\infty} k^2(t)\, dt,$$

We conclude that $E_z \ge E_y$.

5.7 Clearly

$$\overline{\eta(t)} = \widetilde{x(t)} - x(t) \ne 0.$$

5.8   The power spectral density of the noise follows from the transform of (5.6);

$$S_\eta(u) = 2W\,\gamma^2\,\overline{\xi^2}\,[\frac{1}{(2W\gamma)^2}+2\sum_{n=1}^{\infty}\frac{\cos(\pi nu/W)}{n^2+(2W\gamma)^2}]\,\Pi(\frac{u}{2W}).$$

Since [Gradsteyn & Ryzhik #1.445.2]

$$\sum_{k=1}^{\infty}\frac{\cos(kx)}{k^2+a^2}=\frac{\pi}{2a}\frac{\cosh a(\pi-x)}{\sinh(a\pi)}-\frac{1}{2a^2}\quad;0<x\le 2\pi,$$

we conclude that

$$S_\eta(u)=\gamma\pi\overline{\xi^2}\,\frac{\cosh[2\pi\gamma(W-u)]}{\sinh(2\pi\gamma W)}\,\Pi(\frac{u}{2W}).$$

Substituting into (5.9) and evaluating gives

$$\overline{\eta_r^2}/\overline{\xi^2} = \sinh(2\pi\gamma B)\,\coth(2\pi\gamma W)$$
$$= \sinh(2\pi r\gamma W)\,\coth(2\pi\gamma W).$$

A semilog plot of the NINV is shown in Figure A.6 as a function of γW for various values of r.

5.9   By deleting every other sample, the sampling rate parameter becomes $2r$. The NINV for a single sample is $r/(1-r)$. Since

$$2r < r/(1-r)\;;\;\;r<1/2,$$

we conclude that (4.9) yields a smaller NINV.

5.10 Since $x(t)$ is bandlimited, it is analytic. Thus, it's Taylor series expanded about any given point converges everywhere. We choose an interval for which $x(t)$ is identically zero. The resulting Taylor series, however, converges to $x(t) \equiv 0$ everywhere.

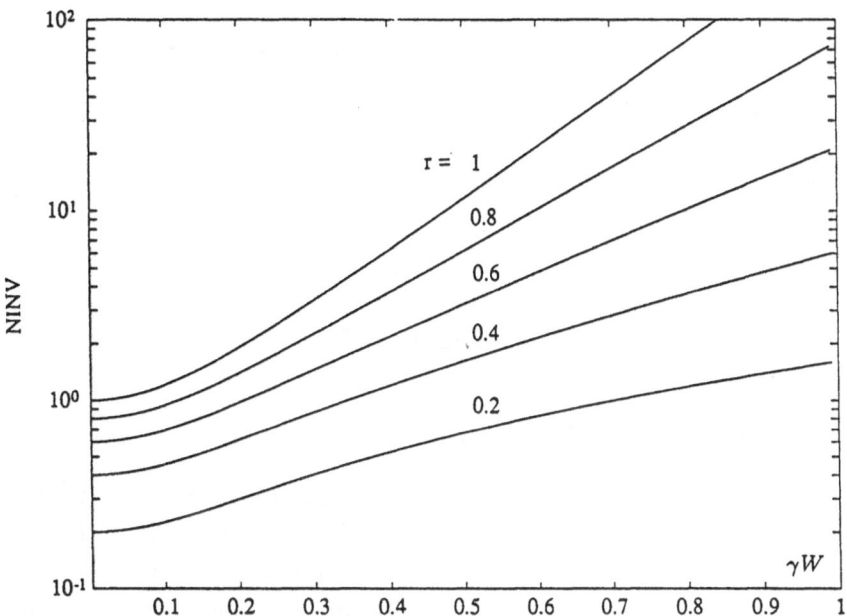

Figure A.6

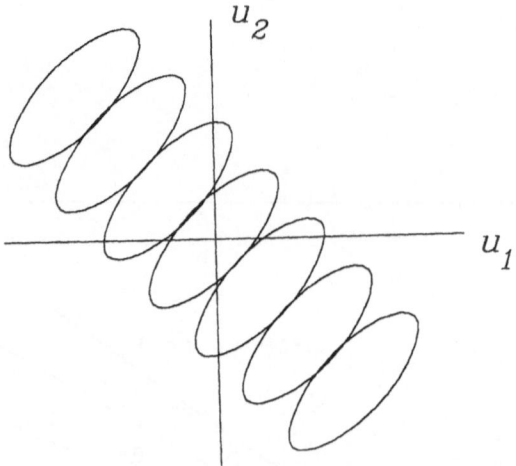

Figure A.7

Solutions, Chapter 6

6.15 (a) The minimum raster sampling rate occurs when sampling with parallel lines with a slope of $9/5$ (or $-9/5$). The resulting minimum sampling rate is equal to the shortest distance between two parallel sides of the parallelogram which can be shown, after a bit of work, to be $1/T = 2\sqrt{56}$.

(b) Use horizontal lines at a sampling interval $T = 1/2$.

(c) Sample with lines of unit slope separated by $T = 1$ intervals. Corresponding spectral replications are shown in Figure A.7.

Solutions, Chapter 7

7.1

$$g_c(t) = f(t)[1 - r_{1-\alpha}(t/T)]$$

$$1 - r_{1-\alpha}(t/T) = \sum_{n=-\infty}^{\infty} c_n \, e^{-j2\pi nt/T}$$

$$\hat{c}_n = \begin{cases} 1 - \alpha & ; \quad n = 0 \\ -(1 - \alpha) \, \text{sinc}[(1 - \alpha)n] & ; \quad n \neq 0 \end{cases}.$$

The b_n coefficients are solutions of

$$\sum_{m=-M}^{M} \hat{b}_m \, \hat{c}_{n-m} = \delta[n] \; ; \; |n| \leq M.$$

And :

$$\hat{\theta}_M(t) = \sum_{m=-M}^{M} \hat{b}_m \, e^{-j2\pi mt}$$

$$\hat{\psi}_M(t) = \hat{\theta}_M(t) \, [1 - r_{1-\alpha}(t/T)].$$

7.2 (a)

$$f(t) = 2\Re \left[\int_0^{\infty} F(u) \, e^{-j2\pi ut} \, du \right].$$

If $f(t)$ is bandlimited, the upper integration limit is B.

(b) From the Figure:

$$F(u) = \frac{c_0 \, G_1(u) - c_1 \, G(u - 1/T)}{c_0^2 - c_{-1} \, c_1}.$$

(c)

$$f(t) = \frac{2}{(c_0^2 - c_{-1}c_1)} \Re[g(t)\{c_0 - c_1 \, e^{j2\pi t/T}\}]$$
$$\quad * [B \, \text{sinc}(Bt) \, e^{j\pi Bt}]$$
$$= \frac{2}{(c_0^2 - c_{-1}c_1)} [\{g(t) \, [c_0 \, \cos \pi Bt$$

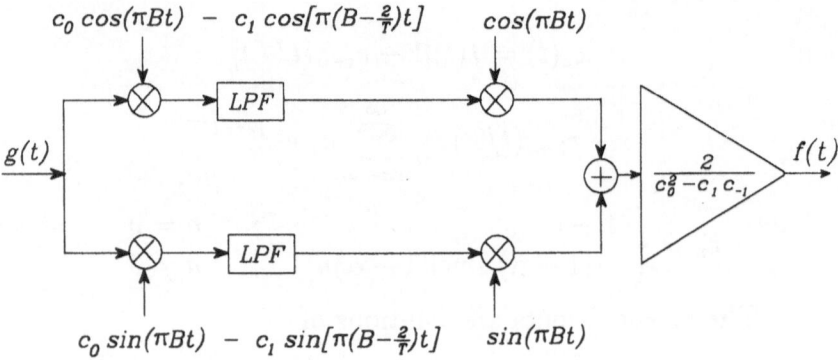

$$c_0 \cos(\pi Bt) \; - \; c_1 \cos[\pi(B-\tfrac{2}{T})t] \qquad \cos(\pi Bt)$$

$g(t)$

LPF

LPF

$$\frac{2}{c_0^2 - c_1 c_{-1}}$$

$f(t)$

$$c_0 \sin(\pi Bt) \; - \; c_1 \sin[\pi(B-\tfrac{2}{T})t] \qquad \sin(\pi Bt)$$

Figure A.8: Restoration of first order aliased data from the shift method illustrated in Fig. 7.27, using Hermetian symmetry. The filters are unity for $|u| < B/2$ and zero elsewhere.

$$\left. -c_1 \cos \pi (B - \frac{2}{T}) t]\} * B\mathrm{sinc}Bt \right| \cos \pi Bt$$

$$+ \frac{2}{(c_0^2 - c_{-1}c_1)} [\{g(t) [c_0 \sin \pi Bt$$

$$\left. -c_1 \sin \pi (B - \frac{2}{T}) t]\} * B\mathrm{sinc}Bt \right| \sin \pi Bt.$$

(d)   See Figure A.8

7.3

$$2B \, \mathrm{sinc}[2B(t - \tau)] = \sum_{n=0}^{\infty} a_n \, \psi_n(t)$$

where

$$a_n \; = \; 2B \int_{-\infty}^{\infty} \mathrm{sinc}[2B(t - \tau)] \, \psi_n(\tau) \, d\tau$$

$$= \; \psi_n(t).$$

Thus:

$$2B \, \mathrm{sinc}[2B(t - \tau)] = \sum_{n=0}^{\infty} \psi_n(t) \psi_n(\tau).$$

7.4 (a) From the power theorem:

$$\int_{-\infty}^{\infty} \text{sinc}(2Bt - n)\, \text{sinc}(2Bt - m)\, dt$$

$$= \int \left[\frac{1}{2B} e^{-j\pi nu/B} \, \Pi(\frac{u}{2B})\right] \left[\frac{1}{2B} e^{-j\pi mu/B} \, \Pi(\frac{u}{2B})\right]$$

$$= \frac{1}{2B}\, \delta[n - m].$$

(b) Clearly:

$$f(t) = \sum_{-\infty}^{\infty} f(\frac{n}{2B})\, \text{sinc}(2Bt - n)$$

$$= \sum_{n=0}^{\infty} a_n\, \psi_n(t).$$

Multiply both sides by $\psi_n(t)$ and integrate. Using (7.30) and (4.47) gives:

$$a_m = \frac{1}{2B} \sum_{n=0}^{\infty} f(\frac{n}{2B})\, \psi_m(\frac{n}{2B}).$$

(c) Here, multiply by $\text{sinc}(2Bt - m)$ and integrate

$$f(m/2B) = 2B \sum_{n=0}^{\infty} a_n\, \psi_n(\frac{m}{2B}).$$

Both are well–posed.

7.5 (a)

$$\int_{-\infty}^{\infty} \psi_n(t)\, \psi_m(t)\, dt$$

$$= 2B \int_{-\frac{T}{2}}^{\frac{T}{2}} \psi_n(t) \int \psi_m(\tau)\, \text{sinc}[2B(t - \tau)]\, d\tau\, dt$$

$$= 2B \int_{-\infty}^{\infty} \psi_m(\tau) \int_{-\frac{T}{2}}^{\frac{T}{2}} \psi_n(t)\, \text{sinc}[2B(t - \tau)]\, dt\, d\tau$$

$$= \lambda_n \int_{-\infty}^{\infty} \psi_n(\tau)\, \psi_m(\tau)\, d\tau$$

$$= \lambda_n\, \delta[n - m].$$

(b) Let

$$I = \int_{-\frac{T}{2}}^{\frac{T}{2}} \psi_n(t)\,\psi_m(t)\,dt$$

$$= \frac{T}{2B} \int_{-B}^{B} \psi_n(\frac{Tu}{2B})\,\psi_m(\frac{Tu}{2B})\,du.$$

$$= \frac{T}{2B}\frac{2B\lambda_n}{T} \int_{-\infty}^{\infty} \psi_n(t)\,\psi_m(t)\,dt$$

Where in the second step we have substituted $t = Tu/2B$. From which (7.33) follows.

7.6 The ambiguity is removed by (7.30) which requires each PSWF to have unit energy.

7.7 (a)

$$\int_{-\infty}^{\infty} |f(t)|^2\,dt = \sum_{n=0}^{\infty} a_n \sum_{m=0}^{\infty} a_m \int_{-\infty}^{\infty} |\psi_n(t)|^2\,dt$$

$$= \sum_{n=0}^{\infty} a_n^2.$$

(b)

$$\int_{-\infty}^{\infty} |h(t)|^2\,dt = \sum_{n=0}^{\infty} \lambda_n\,|b_n|^2.$$

7.8

$$x(t) = x(t)\,\Pi(\frac{t}{T})$$

$$= \sum_{n=0}^{\infty} b_n\,\psi_n(t)$$

$$y(t) = x(t) * 2B\,\text{sinc}(2Bt)$$

$$= \int_{-\infty}^{\infty} x(\tau)\,\text{sinc}[2B(t-\tau)]\,d\tau$$

$$= \sum_{n=0}^{\infty} b_n \int \psi_n(\tau)\,\text{sinc}[(2B(t-\tau)]\,d\tau$$

$$= \sum_{n=0}^{\infty} b_n\,\lambda_n\,\psi_n(t).$$

From (4.20), the energies of x and y are

$$E_x = \sum_{n=0}^{\infty} \lambda_n |b_n|^2 = \sum_{n=0}^{\infty} (\sqrt{\lambda_n}\, b_n)^2$$

and

$$E_y = \sum_{n=0}^{\infty} (\lambda b_n)^2 = \sum_{n=0}^{\infty} \lambda_n (\sqrt{\lambda_n} b_n)^2.$$

We wish to maximize E_y subject to $E_x = 1$. Since λ_0 is the largest eigenvalue, we choose $b_n = \sqrt{\lambda_0}\delta[n]$ and

$$x(t) = \psi_0(t)\, \Pi(\frac{t}{T})/\sqrt{\lambda_0}.$$

The output has energy $E_y = \lambda_0$.

7.9 The results are the same as for extrapolation except that, λ_n becomes $(1 - \lambda_n)$ and $(1 - \mathbf{D}_T)$ replaces $\mathbf{D}_T$. Thus, instead of (7.52), we have:

$$f_N(t) = g_i(t) + \mathbf{D}_T \sum_{n=0}^{\infty} a_n (1 - \lambda_n^N)\psi_N(t)$$

as with the extrapolation case, the validity of this equation can be proven by induction. Convergence again follows due to (7.53).

7.10

$$g_e(t) = \sum_{n=0}^{\infty} g_n\, e^{j2\pi nt/T}$$

$$= \sum_{n=0}^{\infty} a_n\, \psi_n(t) \;;\; |t| < T/2.$$

(a) Multiplying both sides by $\psi_m(t)$ and integrate over $|t| < T/2$:

$$\lambda_m\, a_m = \sum_{n=0}^{\infty} g_n \int_{-\infty}^{\infty} \psi_m(t)\, e^{-j2\pi nt/T}\, dt$$

or, using (7.36):

$$a_m = \frac{1}{\lambda_m} \sum_{n=-\infty}^{\infty} g_n \sqrt{\frac{T\lambda_m}{2B}} \Psi_m(\frac{n}{2B})$$

$$= \sqrt{\frac{T}{2B\lambda_m}} \sum_{n=-\infty}^{\infty} g_n \psi_m(\frac{n}{2B}).$$

The restoration is ill–posed due to the $1/\sqrt{\lambda_m}$ coefficient.

(b) Using the top equation, multiply both sides by $e^{j2\pi nt/T}$ and integrate over $|t| < T/2$:

$$Tg_m = \sum_{n=0}^{\infty} a_n \int_{-\frac{T}{2}}^{\frac{T}{2}} \psi_m(t)\, e^{j2\pi nt/T}\, dt.$$

Again, using (7.36)

$$g_m = \frac{1}{T} \sum_{n=0}^{\infty} a_n \sqrt{\frac{T\lambda_n}{2B}} \psi_m(\frac{-n}{2B})$$

$$= \frac{1}{\sqrt{S}} \sum_{n=0}^{\infty} \sqrt{\lambda_n} a_n \psi_m(\frac{-n}{2B}).$$

The result is well–posed.

7.12 (a) All bandlimited signals are analytic. The only function that can be identically zero over any interval is zero everywhere (Taylor series).

(b) The first two iterations are shown in Figure A.9.

7.13 (a) Instead of (7.42), we have

$$f_{N+1}(t) = g_e(t) + \mathbf{R}(1 - \mathbf{D_T})\mathbf{B_B}\, f_N(t)$$

where the nonlinear half wave rectifier operator is defined by

$$\mathbf{R}\, h(t) = h(t)\, \mu[h(t)].$$

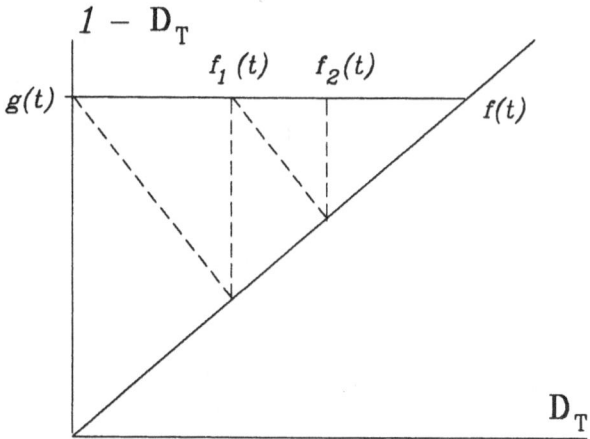

Figure A.9

(b) Here we know $S_\xi(u) \geq 0$. Since

$$\mathbf{B_B} = \mathbf{F}^{-1}\, \mathbf{D_B}\, \mathbf{F}$$

our altered algorithm could be:

$$f_{N+1}(t) = g_e(t) + (1 - \mathbf{D_T})\, \mathbf{F}^{-1}\, \mathbf{R}\, \mathbf{D_B}\, \mathbf{F}\, f_N(t).$$

This is the so–called **minimum negativity constraint** proposed by Howard. Both are special cases of alternated projections onto convex sets [Stark], [Youla and Webb],[Cheung, Marks, and Atlas].

7.14 (a) Let $h(t)$ have finite energy. Then $\mathbf{B_B}\, h(t) = f(t)$ is bandlimited and can be written as

$$f(t) = \sum_{n=0}^{\infty} a_n\, \psi_n(t).$$

(Note that maximum energy occurs when $h(t)$ is chosen to be bandlimited.) Thus,

$$\| \mathbf{H} \| = \sup_{\|f(t)\|=1} \| \mathbf{H}f(t) \|$$

$$\| \mathbf{H} \| = \sup_{\|f(t)\|=1} \| (1 - \mathbf{D_T})f(t) \|$$

where $f(t)$ is bandlimited. Now

$$(1 - \mathbf{D_T})f(t) = \sum_{n=0}^{\infty} a_n (1 - \mathbf{D_T}) \psi_n(t)$$

and

$$\| (1 - \mathbf{D_T})f(t) \|^2 = \sum_{n=0}^{\infty} \sum_{m=0}^{\infty} a_n a_m^* \int_{|t| \geq \frac{T}{2}} \psi_n(t) \psi_m(t)$$

$$= \sum_{n=0}^{\infty} (1 - \lambda_n) | a_n |^2$$

where we have used (5.41). Since, from Exercise 7.7a,

$$\sum_{n=0}^{\infty} | a_n |^2 = 1$$

we choose $a_\infty = 1$ (since $\lambda_\infty = 0$) and $\| \mathbf{H} \| = 1$.

(b)   Same as above, except

$$\| \mathbf{H} \| = \sup_{\|f(t)\|=1} \| \mathbf{D_T} f(t) \| .$$

Since

$$\| \mathbf{D_T} f(t) \|^2 = \sum_{n=0}^{\infty} \lambda_n | a_n |^2,$$

we choose $| a_n | = 1$ since λ_0 is max and $\| \mathbf{H} \| = \sqrt{\lambda_0} < 1$.

7.15

$$x_N(0) = \int_{-\infty}^{\infty} X_N(u)du$$

$$= T \int_{-\infty}^{\infty} \int_{-\infty}^{\infty} \{x_{N-1}(0)\delta(t)$$

$$+ \sum_{n=0}^{\infty} x(nT) \delta(t - \tau)\} e^{-j2\pi ut} \, dt \, du$$

$$= T \int_{-\infty}^{\infty} [x_{N-1}(0) \sum_{n=0}^{\infty} x(nT) \, e^{-j2\pi nut}] du$$

$$= r \, x_{N-1}(0) + r \sum_{n=0}^{\infty} x(nT) \, \text{sinc}(rn)$$

where $r = 2BT$. Letting $N \longrightarrow \infty$ gives

$$x_{\infty}(0) = r x_{\infty}(0) + r \sum_{n=0}^{\infty} x(nT) \, \text{sinc}(rn).$$

Solving for $x_{\infty}(0)$ therefore results in (4.9). Note also that $x_N(t) \longrightarrow x(t)$.

Index